# College
# PHYSICS
*Online*
## for undergraduates
Calculus & non-calculus based

**Part 1** *3rd Edition*

## Eugene de Silva PhD., FRSA

Walter State Community College

**Linus**
Publications, Inc.

Published by Linus Publications, Inc.

Ronkonkoma, NY 11779

ISBN 10: 1-60797-391-X

ISBN 13: 978-1-60797-391-1

Printed in the United States of America.

Print Numbers        5        4        3        2        1

# TABLE OF CONTENTS

## MODULE 8:

# FOREWORD

This book includes some selected notes from my previous books on physics and some selected information freely available online including references to PowerPoint lectures, news articles, and simulated labs. All work, other than my own, is free from copyrights and available to enhance learning of physics. The content copied from creative commons was available under creative commons distribution license. The direct link to that license is, http://creativecommons.org/licenses/by/2.0/

Under this license, all the work used from this site would be subject to Creative Commons Attribution License (CC-BY 2.0) and the credit goes to the authors Sunil Kumar Singh Ph.D. and **Jeffrey W. Schnick**, Ph.D.. Their style of writing and work is similar to the pattern that I wish to use in my online classes. Some concepts from their books have been omitted to thoroughly cover concepts in the time allotment for the online course. You may access the online versions of the books of these two authors directly by visiting the following links:

http://cnx.org/content/col10322/latest/

http://www.anselm.edu/internet/physics/cbphysics/downloadsI.html

This book is an independent book produced by Linus Publications.

My main task in writing this book was to provide students with an easy pathway to study physics from home. Therefore, this book is directly helpful for online physics programs I have written for the Tennessee Board of Regents and other university programs.

In my opinion, it is imperative that we, the educators, change our teaching techniques to accommodate the new trends in social, economic, educational, and political settings. We need to provide our students with educational materials, lectures, and activities with clear objectives to make them see the pathway to success, concise information to enable them to grasp the knowledge in a timely manner, and maximum opportunities to improve and apply their knowledge to real-life situations to see the relevance in what we teach. I hope this physics book encourages the students to study physics further and apply these lessons to daily activities.

I thank my mother Latha for her continuous support, my daughter Eugenie for her unconditional love, and my brother Terrence for his blessings.

My special thanks go to Dr. Jeffrey Horner, the Dean of Natural Science at Walters State Community College for his support and encouragement. Thanks to Dr. David White the Director of online teaching at Walters State Community College for his support from conception to completion in the development of the first online physics course for the Tennessee Board of Regents and to Dr. Celine Santiago-Bass, the Chair of the Science Department at Kaplan University for creating opportunities.

I especially want to thank my former colleagues Howard Titelbaum, Alan Biel, Jack McCann, Jason Robadu, and the late James Green at Lincoln Memorial University, Kennard Lawrence, Sean Cordry, Olena Owen and Douglas Hensley at Walters State Community College, and Thea Leonard at Kaplan University for collegial spirit.

I am most grateful to adjunct professor Betsy Sparks at Walters State Community College for proof-reading this book for its latest version and inclusion of some supplemental problems and solutions.

Finally, I thank Nash Triviani and the team at Linus Publications for the production of this book on time for the classes.

Eugene de Silva

May 15, 2013

Harrogate, TN, U.S.A.

# INTRODUCTION

This book is written for students taking a college physics course either calculus based or non-calculus based.

Each chapter has a set of objectives, lecture notes, an external link to a freely available PowerPoint lecture covering the topic, discussion questions, and an external link to a freely available physics laboratory activity, and student's chapter review. A database of review questions will be available in the future to accompany this book. Instructors may replace the external links with their own PowerPoint lectures and lab activities to match their requirements.

For the laboratory activity, you can use the simulation labs freely provided by PhET, University of Colorado. (link: <a rel="cc:attributionURL" property="cc:attributionName" href="http://phet.colorado.edu">PhET, University of Colorado</a> / <a rel="license" href="http://creativecommons.org/licenses/by/3.0/us/">CC BY3.0</a> ) I am grateful for the work of the authors who have written separate lab activities and the University of Colorado for letting educators and students use this facilities free-of-charge as per the license details. We are also thankful to those academics who have published free PowerPoint lectures on the web.

The fourteen chapters in this book can be completed in fourteen weeks in an online or in class environment. At the end of each chapter, a test should be taken in addition to the two main tests at the end of every seven chapters.

For practice questions and answers, we encourage you to visit, http://www.solvephysics.com/index.html site. This is a free online site copy righted to Copyright © solvephysics.com 2009. All rights reserved. There are several such sites available online, and we encourage students to search these sites in addition to checking the chapter practice quizzes given at the end of this book.

I hope that this easy-to-understand and direct guide in physics would make learning physics more fun and less cumbersome.

# THE SCIENTIFIC METHOD

The scientific method is the single most important aspect a teacher should teach his/her students. Not only can the scientific method be applied to scientific experiments, but also to our daily activities, decision-making, and solving general problems. In our teaching, we should also attempt to improve the following activities within students through the subjects we teach.

These key skills designated by the Department for Education in the United Kingdom are important aspects of the scientific method.

- To improve communication skills
- To improve number skills
- To improve IT skills
- To learn to work with others
- To improve own learning and performance
- To solve problems

The general steps involved in scientific method are as follows:

1. Name the problem or question.
2. Form an educated guess (hypothesis) of the cause of the problem and make predictions based upon the hypothesis.
3. Test your hypothesis by doing an experiment or study with proper controls.
4. Check and interpret your results
5. Report your results.

For more information on scientific method, you may visit

http://www.sciencebuddies.com/mentoring/project_scientific_method.shtml.

# SOLVING PROBLEMS IN PHYSICS

To solve problems, you may want to implement the GUESS, or when appropriate, the GUPPESS method as follows:

| Givens | Givens |
|---|---|
| Unknowns | Unknowns |
| Equation | Principle |
| Substitution | Picture |
| Solution | Substitution |
| | Solution |

It is also a good idea to incorporate these steps into any problem that your teacher needs you to evaluate. It makes working through your thought processes easier.

e.g:

A car accelerates at a rate of 0.60 m/s². How long does it take the car to go from 55 mi/hr to 60 mi/hr?

**GIVENS:** $a = 0.60$ m/s²; $v_0 = 55$ mi/hr; v or $v_f = 60$ mi/hr

**UNKNOWNS:** t=? seconds

**PRINCIPLE:** You could use either of at least two equations to find the time when given acceleration and beginning and ending velocities: $a = \dfrac{\Delta v}{t}$ or $v = v_0 + at$ Part of the problem is that you need to get from mi/hr to m/s – and for that you use dimensional analysis

**PICTURE:** Probably not that helpful in this instance, but you can draw one.

**EQUATION, SUBSTITUTION, and SOLUTION:**

$v = v_0 + at$

$60$ mi/hr $= 55$ mi/hr $+ (0.60$ m/s²$)t$

$60$ mi/hr $- 55$ mi/hr $= (0.60$ m/s²$)t$

$5$ mi/hr $= (0.60$ m/s²$)t$

Here is where you may want to use dimensional analysis in a side bar to convert 5 mi/hr into m/s².

$$\frac{5mi}{1hr} \times \frac{1609m}{1mi} \times \frac{1hr}{3600s} = \frac{2.2m}{1s} = 2.2m/s$$

Now substitute 2.2 m/s into the previous equation where you had 5 mi/hr.

$5$ mi/hr $= (0.60$ m/s²$)t$

$2.2 \text{ m/s} = (0.60 \text{ m/s}^2)t$

$\dfrac{2.2 \text{m/s}}{(0.60 \text{m/s}^2)} = t$  NOTE: Follow the units! -> $\dfrac{2.2 \text{m}}{1 \text{s}} \times \dfrac{1 \text{s}^2}{0.60 \text{m}}$

**3.7 s = t**

# Some Mathematical formulae

## Trigonometry

Trigonometry will become important when you study vectors.

$$\sin \theta = \frac{opposite}{hypotenuse}$$

$$\tan \theta = \frac{opposite}{adjacent}$$

$$\cos \theta = \frac{adjacent}{hypotenuse}$$

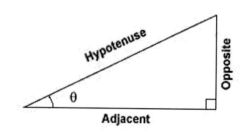

## Error Percentage Calculations

$$\% \, Change = \frac{change}{original} \, 100\%$$

$$\% \, Change = \frac{new - original}{original} \, 100\%$$

## Quadratic Formula

$$ax^2 + bx + c = 0$$

$x$ is the variable and $a$, $b$, and $c$ are constants

The x value is given by the **quadratic formula**:

$$x = \frac{-b \pm \sqrt{b^2 - 4ac}}{2a}$$

# Calculus Formulae

Trigonometric Formulae

1. $\sin^2\theta + \cos^2\theta = 1$

2. $1 + \tan^2\theta = \sec^2\theta$

3. $1 + \cot^2\theta = \csc^2\theta$

4. $\sin(-\theta) = -\sin\theta$

5. $\cos(-\theta) = \cos\theta$

6. $\tan(-\theta) = -\tan\theta$

7. $\sin(A + B) = \sin A\cos B + \sin B\cos A$

8. $\sin(A - B) = \sin A\cos B - \sin B\cos A$

9. $\cos(A + B) = \cos A\cos B - \sin A\sin B$

10. $\cos(A - B) = \cos A\cos B + \sin A\sin B$

11. $\sin 2\theta = 2\sin\theta\cos\theta$

12. $\cos 2\theta = \cos^2\theta - \sin^2\theta = 2\cos^2\theta - 1 = 1 - 2\sin^2\theta$

13. $\tan\theta = \dfrac{\sin\theta}{\cos\theta} = \dfrac{1}{\cot\theta}$

14. $\cot\theta = \dfrac{\cos\theta}{\sin\theta} = \dfrac{1}{\tan\theta}$

15. $\sec\theta = \dfrac{1}{\cos\theta}$

16. $\csc\theta = \dfrac{1}{\sin\theta}$

17. $\cos(\dfrac{\pi}{2} - \theta) = \sin\theta$

18. $\sin(\dfrac{\pi}{2} - \theta) = \cos\theta$

## *Differentiation Formulas*

1. $\dfrac{d}{dx}(x^n) = nx^{n-1}$

2. $\dfrac{d}{dx}(fg) = fg' + gf'$

3. $\dfrac{d}{dx}(\dfrac{f}{g}) = \dfrac{gf' - fg'}{g^2}$

4. $\dfrac{d}{dx} f(g(x)) = f'(g(x))g'(x)$

5. $\dfrac{d}{dx}(\sin x) = \cos x$

6. $\dfrac{d}{dx}(\cos x) = -\sin x$

7. $\dfrac{d}{dx}(\tan x) = \sec^2 x$

8. $\dfrac{d}{dx}(\cot x) = -\csc^2 x$

9. $\dfrac{d}{dx}(\sec x) = \sec x \tan x$

10. $\dfrac{d}{dx}(\csc x) = -\csc x \cot x$

11. $\dfrac{d}{dx}(e^x) = e^x$

12. $\dfrac{d}{dx}(a^x) = a^x \ln a$

13. $\dfrac{d}{dx}(\ln x) = \dfrac{1}{x}$

14. $\dfrac{d}{dx}(Arc\sin x) = \dfrac{1}{\sqrt{1-x^2}}$

15. $\dfrac{d}{dx}(Arc\tan x) = \dfrac{1}{1+x^2}$

16. $\dfrac{d}{dx}(Arc\sec x) = \dfrac{1}{|x|\sqrt{x^2-1}}$

17. $\dfrac{dy}{dx} = \dfrac{dy}{dx} \times \dfrac{du}{dx}$    Chain Rule

## Integration Formulae

1. $\int a\, dx = ax + C$

2. $\int x^n\,dx = \dfrac{x^{n+1}}{n+1} + C, \quad n \neq -1$

3. $\int \dfrac{1}{x}\,dx = \ln|x| + C$

4. $\int e^x\,dx = e^x + C$

5. $\int a^x\,dx = \dfrac{a^x}{\ln a} + C$

6. $\int \ln x\,dx = x\ln x - x + C$

7. $\int \sin x\,dx = -\cos x + C$

8. $\int \cos x\,dx = \sin x + C$

9. $\int \tan x\,dx = \ln|\sec x| + C \ \text{ or } \ -\ln|\cos x| + C$

10. $\int \cot x\,dx = \ln|\sin x| + C$

11. $\int \sec x\,dx = \ln|\sec x + \tan x| + C$

12. $\int \csc x\,dx = \ln|\csc x - \cot x| + C$

13. $\int \sec^2 x\,dx = \tan x + C$

14. $\int \sec x \tan x\,dx = \sec x + C$

15. $\int \csc^2 x\,dx = -\cot x + C$

16. $\int \csc x \cot x\,dx = -\csc x + C$

17. $\int \tan^2 x\,dx = \tan x - x + C$

18. $\int \dfrac{dx}{a^2 + x^2} = \dfrac{1}{a}\,Arc\tan\left(\dfrac{x}{a}\right) + C$

19. $\int \dfrac{dx}{\sqrt{a^2 - x^2}} = Arc\sin\left(\dfrac{x}{a}\right) + C$

20. $\int \dfrac{dx}{x\sqrt{x^2 - a^2}} = \dfrac{1}{a}\,Arc\sec\dfrac{|x|}{a} + C = \dfrac{1}{a}\,Arc\cos\left|\dfrac{a}{x}\right| + C$

# Formulae and Theorems

<u>Definition of Limit</u>: Let $f$ be a function defined on an open interval containing $c$ (except possibly at $c$) and let $L$ be a real number. Then $\lim\limits_{x \to a} f(x) = L$ means that for each $\varepsilon > 0$ there exists a $\delta > 0$ such that $|f(x) - L| < \varepsilon$ whenever $0 < |x - c| < \delta$.

1b. A function $y = f(x)$ is <u>continuous</u> at $x = a$ if

i). $f(a)$ exists

ii). $\lim\limits_{x \to a} f(x)$ exists

iii). $\lim\limits_{x \to a} = f(a)$

## Even and Odd Functions

1.  A function $y = f(x)$ is <u>even</u> if $f(-x) = f(x)$ for every $x$ in the function's domain.

    Every even function is symmetric about the y-axis.

2.  A function $y = f(x)$ is <u>odd</u> if $f(-x) = -f(x)$ for every $x$ in the function's domain.

    Every odd function is symmetric about the origin.

## Periodicity

A function $f(x)$ is periodic with period $p\,(p > 0)$ if $f(x + p) = f(x)$ for every value of $x$.

<u>Note</u>: The period of the function $y = A\sin(Bx + C)$ or $y = A\cos(Bx + C)$ is $\dfrac{2\pi}{|B|}$.

The amplitude is $|A|$. The period of $y = \tan x$ is $\pi$.

## Intermediate-Value Theorem

A function $y = f(x)$ that is continuous on a closed interval $[a, b]$ takes on every value between $f(a)$ and $f(b)$.

<u>Note</u>: If $f$ is continuous on $[a, b]$ and $f(a)$ and $f(b)$ differ in sign, then the equation

$f(x) = 0$ has at least one solution in the open interval $(a, b)$.

## Limits of Rational Functions as $x \to \pm\infty$

i.  $\lim\limits_{x \to \pm\infty} \dfrac{f(x)}{g(x)} = 0$ if the degree of $f(x) <$ the degree of $g(x)$

*Example:* $\lim\limits_{x \to \infty} \dfrac{x^2 - 2x}{x^3 + 3} = 0$

ii. $\lim\limits_{x \to \pm\infty} \dfrac{f(x)}{g(x)}$ is infinite if the degrees of $f(x) >$ the degree of $g(x)$

*Example:* $\lim\limits_{x \to \infty} \dfrac{x^3 + 2x}{x^2 - 8} = \infty$

iii. $\lim\limits_{x \to \pm\infty} \dfrac{f(x)}{g(x)}$ is finite if the degree of $f(x) =$ the degree of $g(x)$

*Example:* $\lim\limits_{x \to \infty} \dfrac{2x^2 - 3x + 2}{10x - 5x^2} = -\dfrac{2}{5}$

# Horizontal and Vertical Asymptotes

1.  A line $y = b$ is a <u>horizontal asymptote</u> of the graph $y = f(x)$ if either

    $\lim\limits_{x \to \infty} f(x) = b$ or $\lim\limits_{x \to -\infty} f(x) = b$.

2.  A line $x = a$ is a <u>vertical asymptote</u> of the graph $y = f(x)$ if either

    $\lim\limits_{x \to a^+} f(x) = \pm\infty$ or $\lim\limits_{x \to a^-} = \pm\infty$.

# Average and Instantaneous Rate of Change

i.  <u>*Average Rate of Change*</u>: If $(x_0, y_0)$ and $(x_1, y_1)$ are points on the graph of $y = f(x)$, then the average rate of change of $y$ with respect to $x$ over the interval $[x_0, x_1]$ is

    $\dfrac{f(x_1) - f(x_0)}{x_1 - x_0} = \dfrac{y_1 - y_0}{x_1 - x_0} = \dfrac{\Delta y}{\Delta x}$.

ii. <u>*Instantaneous Rate of Change*</u>: If $(x_0, y_0)$ is a point on the graph of $y = f(x)$, then the instantaneous rate of change of $y$ with respect to $x$ at $x_0$ is $f'(x_0)$.

    $f'(x) = \lim\limits_{h \to 0} \dfrac{f(x+h) - f(x)}{h}$

# The Number *e* as a limit

i.  $\lim\limits_{n \to +\infty} \left(1 + \dfrac{1}{n}\right)^n = e$

ii. $\lim_{n \to 0} \left(1 + \dfrac{n}{1}\right)^{\frac{1}{n}} = e$

## Rolle's Theorem

If $f$ is continuous on $[a,b]$ and differentiable on $(a,b)$ such that $f(a) = f(b)$, then there is at least one number $c$ in the open interval $(a,b)$ such that $f'(c) = 0$.

## Mean Value Theorem

If $f$ is continuous on $[a,b]$ and differentiable on $(a,b)$, then there is at least one number $c$ in $(a,b)$ such that $\dfrac{f(b) - f(a)}{b - a} = f'(c)$.

## Extreme-Value Theorem

If $f$ is continuous on a closed interval $[a,b]$, then $f(x)$ has both a maximum and minimum on $[a,b]$.

To find the maximum and minimum values of a function $y = f(x)$, locate

1.  the points where $f'(x)$ is zero <u>or</u> where $f'(x)$ fails to exist.

2.  the end points, if any, on the domain of $f(x)$.

<u>Note</u>: These are the <u>only</u> candidates for the value of $x$ where $f(x)$ may have a maximum or a minimum.

Let $f$ be differentiable for $a < x < b$ and continuous for a $a \le x \le b$,

1.  If $f'(x) > 0$ for every $x$ in $(a,b)$, then $f$ is increasing on $[a,b]$.

2.  If $f'(x) < 0$ for every $x$ in $(a,b)$, then $f$ is decreasing on $[a,b]$.

# Errors

Measurement is the basis of scientific study. All measurements are, however, approximate values (not true values) within the limitation of a measuring device, measuring environment, process of measurement and human error. We seek to minimize uncertainty and hence error to the extent possible.

Further, there is an important aspect of reporting measurement. It should be consistent, systematic and revealing in the context of accuracy and precision. We must understand that an error in basic quantities propagate through mathematical formula leading to compounding of errors and misrepresentation of quantities.

Errors are broadly classified into two categories:

- Systematic error

- Random error

A systematic error impacts "accuracy" of the measurement. Accuracy means how close is the measurement with respect to "true" value. A "true" value of a quantity is a measurement, when errors on all accounts are minimized. We should distinguish "accuracy" of measurement with "precision" of measurement, which is related to the ability of an instrument to measure values with greater details (divisions).

The measurement of a weight on a scale with marking in kg is 79 kg, whereas measurement of the same weight on a different scale having further divisions in hectogram is 79.3 kg. The later weighing scale is more precise. The precision of measurement of an instrument, therefore, is a function of the ability of an instrument to read smaller divisions of a quantity.

**Summary:**

1. True value of a quantity is an "unknown". We cannot know the true value of a quantity, even if we have measured it by chance, as we do not know the exact value of error in measurement. We can only approximate true value with greater accuracy and precision.

2. An accepted "true" measurement of a quantity is a measurement, when errors on all accounts are minimized.

3. "Accuracy" means how close the measurement is with respect to "true" measurement. It is associated with systematic error.

4. "Precision" of measurement is related to the ability of an instrument to measure values in greater detail. It is associated with random error.

# Systematic error

A systematic error results due to faulty measurement practices. An error of this type is characterized by deviation in one direction from the true value. Itt means that the error is introduced, which is either less than or greater than the true value. Systematic error impacts the accuracy of measurement – not the precision of the measurement.

Systematic error results from:

1. faulty instrument

2. faulty measuring process and

3. personal bias

Clearly, this type of error cannot be minimized or reduced by repeated measurements. A faulty machine, for example, will not improve accuracy of measurement by repeating measurements.

# Instrument error

A zero error, for example, is an instrument error, which is introduced in the measurement consistently in one direction. A zero error results when the zero mark of the scale does not match with pointer. We can realize this with the weighing instrument (scale) we use in our home. Often, the pointer is off the zero mark of the scale. Moreover, the scale may in itself be not uniformly marked or may not be properly calibrated. In vernier calipers, the nine divisions of main scale should be exactly equal to ten divisions of vernier scale. In a nutshell, we can say that the instrument error occurs due to faulty design of the instrument. We can minimize this error by replacing the instrument or by making a change in the design of the instrument.

## Procedural error

A faulty measuring process may include an inappropriate physical environment, procedural mistakes and a lack of understanding of the process of measurement. For example, if we are studying the magnetic effect of current, then it would be erroneous to conduct the experiment in a place where strong currents are flowing nearby. Similarly, while taking the temperature of a human body, it is important to know which of the human parts is more representative of body temperature.

This type of error can be minimized by periodically assessing the measurement process and improvising the system. Consulting a subject expert or simply conducting an audit of the measuring process is also helpful in light of new facts and advancements.

## Personal bias

A personal bias is introduced by human habits, which are not conducive for accurate measurement. Consider for example, the reading habit of a person. He or she may have the habit of reading scales from an inappropriate distance and from an oblique direction. The measurement, therefore, includes error because of parallax.

**Parallax**

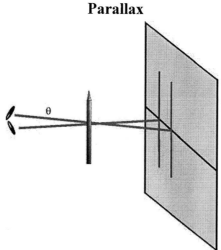

**Figure 1: The position of pencil changes with respect to a mark on the background.**

We can appreciate the importance of parallax by just holding a finger (pencil) in the hand, which is stretched horizontally. We keep the finger in front of our eyes against some reference marking in the background. Now, we look at the finger by closing one eye at a time and note the relative displacement of the finger with respect to the mark in the static background. We can do this experiment any time as shown in the figure above. The parallax results due to the angle at which we look at the object.

It is important that we read position of a pointer or a needle on a scale normally to avoid error because of parallax.

**Parallax**

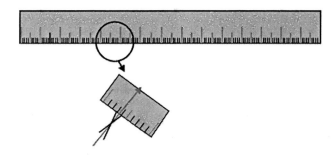

**Figure 2: Parallax error is introduced as we may read values at an angle.**

# Random errors

Random error unlike systematic error is not unidirectional. Some of the measured values are greater than true value; some are less than true value. The errors introduced are sometimes positive and sometimes negative with respect to true value. It is possible to minimize this type of error by repeating measurements and applying statistical techniques to get closer value to the true value.

Another distinguishing aspect of random error is that it is not biased. It is there because of the limitation of the instrument in hand and the limitation on the part of human ability. No human being can repeat an action in exactly the same manner. Hence, it is likely that the same person reports different values with the same instrument, which measures the quantity correctly.

# Least count error

Least count error results due to the inadequacy of resolution of the instrument. We can understand this in the context of least count of a measuring device. The least count of a device is equal to the smallest division on the scale. Consider the meter scale that we use. What is its least count? Its smallest division is in millimeter (mm). Hence, its least count is 1 mm i.e. $10^{-3}$ m i.e. 0.001 m. Clearly, this meter scale can be used to measure length from $10^{-3}$ m to 1 m. It is worth to know that least count of a vernier scale is $10^{-4}$ m and that of screw gauge and spherometer is $10^{-5}$ m.

Returning to the meter scale, we have the dilemma of limiting ourselves to the exact measurement up to the precision of marking or should be limited to a step before. For example, let us read the measurement of a piece of a given rod. One end of the rod exactly matches with the zero of scale. The other end lies at the smallest markings at 0.477 m (= 47.7 cm = 477 mm). We may argue that measurement should be limited to the marking, which can be definitely relied. If so, then we would report the length as 0.47 m, because we may not be definite about millimeter reading.

This is, however, unacceptable, as we are sure that length consists of some additional length – only thing that we may err as the reading might be 0.476 m or 0.478 m instead of 0.477 m. There is a definite chance of error due to limitation in reading such small divisions. We would be more precise and accurate by reporting measurement as 0.477 ± some agreed level of anticipated error. Generally, the accepted level of error in reading the smallest division is considered to be half of the least count. Hence, the reading would be:

$$\Rightarrow x = 0.477 \pm 0.0005 m$$

If we report the measurement in centimeter,

$\Rightarrow x = 47.7 \pm 0.05 cm$

If we report the measurement in millimeter,

$\Rightarrow x = 477 \pm 0.5 mm$

# Mean value of measurements

It has been pointed out that random error, including that of least count error, can be minimized by repeating measurements. It is so because errors are not unidirectional. If we take average of the measurements from the repeated measurements, it is likely that we minimize error by canceling out errors in opposite directions.

Here, we are implicitly assuming that measurement is free of "systematic errors". The averaging of the repeated measurements, therefore, gives the best estimate of "true" value. As such, average or mean value ( $a_m$ ) of the measurements (excluding "of- beat" measurements) is the notional "true" value of the quantity being measured. In fact, it is reported as true value, being our best estimate.

# Error Propagation

In this module, we shall introduce some statistical analysis techniques to improve our understanding about error and enable reporting of error in the measurement of a quantity. There are three related approaches, which involves measurement of:

- Absolute error

- Relative error

- Percentage error

- Absolute error

The absolute error is the magnitude of error as determined from the difference of measured value from the mean value of the quantity. The important thing to note here is that absolute error is concerned with the magnitude of error – not the direction of error. For a particular nth measurement,

$$|\Delta x_n| = |x_n - x_n| \, |$$

where "$x_m$" is the mean or average value of measurements and "$x_n$" is the nth instant of measurement.

In order to calculate a few absolute values, we consider a set of measured data for the length of a given rod. Note that we are reporting measurements in centimeter.

$x_1 = 47.7 \, cm, \; x_2 = 47.5 \, cm, \; x_3 = 47.8 \, cm, \; x_4 = 47.4 \, cm$ and $x_5 = 47.7 \, cm$

The mean value of length is,

$\Rightarrow x_m = 47.62 \, cm$

It is evident from the individual values that the least count of the scale (smallest division) is 0.001 m = 0.1 cm. For this reason, we limit mean value to the first decimal place. Hence, we round off the last but one digit as:

$x_m = 47.6$ cm

This is the mean or true value of the length of the rod. Now, absolute error of each of the five measurements are,

$|\Delta x_1| = |x_m - x_1| = |47.6 - 47.7| = |-0.1 = .1$ cm

$|\Delta x_2| = |x_m - x_2| = |47.6 - 47.5| = |0.1| = 0.1$ cm

$|\Delta x_3| = |x_m - x_3| = |47.6 - 47.8| = |-0.2| = 0.2$ cm

$|\Delta x_4| = |x_m - x_4| = |47.6 - 47.4| = |0.2| = 0.2$ cm

$|\Delta x_5| = |x_m - x_5| = |47.6 - 47.7| = |-0.1| = 0.1$ cm

## Mean absolute error

Earlier, it was stated that a quantity is measured with a range of error specified by half the least count. This is a generally accepted range of error. Here, we shall work to calculate the range of the error, based on the actual measurements and not go by any predefined range of error as that of generally accepted range of error. This means that we want to determine the range of error, which is based on the deviations in the reading from the mean value.

Absolute error associated with each measurement tells us how far the measurement can be off the mean value. The absolute errors calculated, however, may be different. Now the question is, which of the absolute errors should be taken for our consideration? We take the average of the absolute error:

The value of measurement, now, will be reported with the range of error as:

$x = x_m \pm \Delta x_m$

Extending this concept of defining range to the earlier example, we have,

$\Rightarrow \Delta x_m = 0.1$ cm.

We should note here that we have rounded the result to reflect that the error value that has same precision as that of measured value. The value of the measurement with the range of error, then, is :

$\Rightarrow x = 47.6 \pm 0.1$ cm

A plain reading of above expression is "the length of the rod lies in between 47.5 cm and 47.7 cm". For all practical purpose, we shall use the value of x = 47.6 cm with caution, in that this quantity involves an error of the magnitude of "0.1 cm" in either direction.

For more details on other types of errors, please visit:

http://www.lon-capa.org/~mmp/labs/error/e2.htm#V

# RODP Physics 2010 and RODP Physics 2110

This book matches the modules and chapters of RODP online physics 2010 order of lessons. For RODP online physics 2110 course, the following content order is given, although it is advised that the students follow the headings and content rather than the chapter order. This means reading the information within the book in any order, and also checking on lecture links given under useful websites on the content page of the course homepage.

| RODP Physics 2010 Module (Chapters follow the order) | RODP Physics 2110 Module |
|---|---|
| Chapter 1: Units, Physical Quantities, and Vectors | Units, Vectors, and Motion in one, two, and three dimensions (Read chapters 1, 2, and 3 of the book) |
| Chapter 2: Motion in one Dimension | The Laws of Motion (Read chapter 4) |
| Chapter 3: Vectors and Two Dimensional Motions | Work and Kinetic Energy (Read chapter 5) |
| Chapter 4: The Laws of Motion | Potential Energy and Energy Conservation (Read chapter 5) |
| Chapter 5: Energy | Momentum, Impulse, and Collisions and Rotation of Rigid Bodies (Read chapter 6) |
| Chapter 6: Momentum and Collisions | Dynamics of Rotational Motion (Read chapters 7 and 8) |
| Chapter 7: Rotational Motion and the Law of Gravity | Equilibrium and Elasticity, Gravitation, and Periodic Motion (Read chapters 7 and 8) |
| Chapter 8: Rotational Equilibrium and Rotational Dynamics | Fluid Mechanics and Sound (Read chapters 9, 13, and 14) |
| Chapter 9: Solids and Fluids | |
| Chapter 10: Thermal Physics | |
| Chapter 11: Energy in Thermal Process | |
| Chapter 12: The Laws of Thermodynamics | |
| Chapter 13: Vibrations and Waves | |
| Chapter 14: Sound | |

# Module 1

## CHAPTER-1

# Units, Physical Quantities, and Vectors

## Objectives

At the end of this lesson, you should be able to:

1. Express uncertainty in measurements and significant figures;

2. Apply dimensional analysis;

3. Apply correct units;

4. Describe and solve problems involving vectors.

## Lecture Notes

- All measurements must use a standard or a unit.

- These measurements are used to compare the quantity or amount of various substances.

- All physical quantities have a number and a unit.

- In all scientific work, a standardized system is used. This is known as System International or SI.

- SI units are systematic, coherent, and convenient in application.

- In the SI system, seven quantities are selected as basic physical quantities (Table 1.1).

- A basic physical quantity does not depend on any other quantity. An SI unit is given for each property.

- A derived physical quantity is one that is derived by combining one or more of the basic physical quantities (Table 1.2).

- At times, the numerical value of a quantity may have to be expressed in fractions of SI units. Such prefixes are given in Table 1.3.

### Table 1.1  Basic Physical Quantities

| Basic Physical Quantity | Symbol for unit | S.I. Unit |
|---|---|---|
| Length ( l ) | m | meter |
| Mass  ( m or w) | kg | kilogram |
| Time ( t ) | s | second |
| Electric Current ( I ) | A | ampere |
| Temperature ( T ) | K | Kelvin |
| Amount of Substance ( n ) | mol | mole |
| Luminous Intensity ( Iv ) | cd | candela |

### Table 1.2  Derived Physical Quantities

| Physical Quantity | Derivation | SI unit (other unit) |
|---|---|---|
| Area   (a) | $a = l \times l$ | $m^2$ |
| Volume (V) | $V = l \times l \times l$ | $m^3$ |
| Molar mass (M) | $M = m/n$ | $kg\ mol^{-1}$ |
| Density (d) | $d = m/v$ | $kg\ m^{-3}$ |
| Molality (m) | $m = n/m_{solvent}$ | $mol\ kg^{-1}$ |
| Concentration ( C) | $C = n/v$ | $mol\ m^{-3}$ |
| Acceleration (a) | $a = v/t$ | $m\ s^{-2}$ |
| Velocity (v) | $v = l/t$ | $m\ s^{-1}$ |
| Force (F) | $F = m \times a$ | $kg\ m\ s^{-2}$ (Newton )(N) |
| Pressure (P) | $P = F/a$ | $kg\ m^{-1}\ s^{-2}$ (Pascal ) (Pa) |
| Energy (E) | $E = F \times l$ | $kg\ m^2 s^{-2}$ (Joule) (J) |

| Power (W) | W = E/t | kg m$^2$ s$^{-3}$ |
| | | (Watt) (W) |
| | | (Js$^{-1}$) |
| Electrical Potential (V) | V = W/I | kg m$^2$s$^{-3}$A$^{-1}$ |
| | | (Volt) (V) |
| | | (WA$^{-1}$) (JC$^{-1}$) |
| Electrical Charge (Q) | Q = I x t | A s |
| | | (Coulomb) (C) |
| Electrical Resistance (R ) | R = V/I | kg m$^2$ s$^{-3}$ A$^{-2}$ |
| Electrical Conductance ( G ) | G = I/R | A$^3$ kg$^{-1}$ m$^{-2}$s$^3$ |
| | | (Siemens) (S) |
| | | ( A $\Omega^{-1}$) |

**Table 1.3    Fractions and Multiples of SI units**

| Multiple | Symbol | Prefix name | its value |
|---|---|---|---|
| 10$^9$ | G | giga- | one billion times |
| 10$^6$ | M | mega- | one million times |
| 10$^3$ | k | kilo- | one thousand times |
| 10$^2$ | h | hecto- | one hundred times |
| 10$^1$ | da | deka- | ten times |
| 10$^{-1}$ | d | deci- | one-tenth of |
| 10$^{-2}$ | c | centi- | one-hundredth of |
| 10$^{-3}$ | m | milli- | one thousandth of |
| 10$^{-6}$ | μ | micro- | one-millionth of |
| 10$^{-9}$ | n | nano- | one-billionth of |
| 10$^{-12}$ | p | pico- | one-trillionth of |

The above prefixes could be followed by any of the base unit names. Base units are grams, meters, liters, seconds, etc.

**Table 1.4 SI units compared to English units**

| Quantity | SI unit | related units | English units |
|---|---|---|---|
| Length | meter (m) | cm or mm | foot (ft) |
| Mass | kilogram (kg) | grams or mg | pound (lb) |
| Volume | cubic meter (m³) | liters or ml | quart (qt) |
| Temperature | kelvin (K) | Celsius (°C) | Fahrenheit (°F) |
| Pressure | pascal (Pa) | torr or atm. | lb per sq. inch (psi) |
| Energy | joule (J) | calorie(cal) | British thermal unit (Btu) |

# Conversions:

Conversion from one unit to another can be done by following the factor label method given below:

1. Write the given number and unit.

2. Set up a conversion factor, which is a fraction used to convert one unit to another.

   This is done as follows:

   i.   Place the given unit as the denominator of conversion factor.

   ii.  Place the desired unit as numerator.

   iii. Place a "1" in front of the larger unit.

   iv.  Determine the number of smaller units needed to make "1" of the larger unit.

3. Cancel units and solve the problem.

   e.g.

   10 mm =    m

   $$10 \, \text{mm} \quad \frac{1m}{1000mm} \quad \frac{1}{100}$$

   9 dal =  dl

   $$9 \, \text{dal} \quad \frac{10 \, l}{1 \, dal} X \frac{10 \, dl}{1 \, l} = 900 \, dl$$

   Some useful conversions:
   * Relationship between cm³        ³

   $$cm^3 \qquad ^{-2})^3 = 10^{-6}m^3$$
   $$dm^3 = (m \times 10^{-1})^3 = 10^{-3}m^3$$
   $$cm^3 = 10^{-3}dm^3$$

$1000cm^3$     $^3cm^3=10^3x10^{-3}dm^3$

$1000cm^3$       $3$

$1000cm^3 = 1$ liter $= 1dm^3$

- A mole ( symbol: mol) is used to describe the amount of a substance.

- 1 mole contains $6.023x10^{23}$ particles of the substances (or elementary units).

e.g.

1. 1 mole of 12 C (carbon 12 isotope) contains $6.023x10^{23}$ atoms of 12 C in 12g of carbon.

2. Similarly, one mole of $O_2$ has $6.023x10^{23}$ molecules in 32g of weight. The number of atoms in one mole will be ($6.023x10^{23}x2$ ), since an $O_2$ molecule has two atoms.

The following formula is used to calculate the molar mass:

M/n = M

M= molar mass

n= Number of moles

m= mass

If M is for a molecular weight, then it is known as the relative molecular mass (Mr).  If M is for an atomic weight, then it is known as the relative atomic mass (Ar).

## Table 1.5 Conversion Factors

| English System | English System | Metric System | Metric System |
|---|---|---|---|
| One pound    = | 16 ounces    = | 0.4536 Kg  = | 453.6 grams |
| One Foot    = | 12 inches    = | 0.3048 m   = | 30.48 cm |
| One yard   = | 36 inches   = | 0.914 m   = | 91.44 cm |
| One mile   = | 5,280 feet   = | 1.609 Km   = | 1609 m |
| One gallon   = | 4.0 quarts   = | 3.785 liters  = | $3.785 x 10^3$ ml |

**Table 1.6  Other Conversion Factors**

| | | |
|---|---|---|
| Mass units | One ton = | 2,000 lbs |
| Mass units | 1 amu = | $1.6606 \times 10^{-24}$ grams |
| Length units | One Angstrom (A) = | $1.0 \times 10^{-10}$ m |
| Volume units | One pint = | 16 fluid ounces |
| Time units | One hour = | 3600 seconds |
| Temperature units | Zero Celsius = | 273.15 kelvin |
| Energy units | One calorie = | 4.184 Joules |

# Exponential Numbers (scientific or exponential notation):

One uses exponential numbers when dealing with very large numbers (greater than one thousand) or very small numbers (less than one tenth).

Scientific notations take the form of $M \times 10^n$. When n is positive, the number in the standard form is a larger number; when n is negative, the number is between zero and one.

Let us express the number 80,000. in scientific notation, using exponential numbers. By moving the decimal four places to the left, we have the number $8.0 \times 10^4$. This is 80,000. expressed in scientific notation (or exponential notation). A rule of thumb is to express the number between one and ten and raise it to the necessary exponential value. e.g.

Convert 1,700,000 to scientific notation.

The decimal point is placed to get only one digit to the left of the decimal.

The answer is $1.7 \times 10^6$

Convert .0000036 to scientific notation.

The answer is $3.6 \times 10^{-6}$

One must be able to multiply, divide, add and subtract using exponential numbers.  We will discuss each conversion individually.

To multiple numbers expressed in scientific notation, simply add the exponent values and multiply the number expressed between one and ten. For example:

Multiply $3.0 \times 10^6$ times $2.1 \times 10^3$ .  First multiply $3.0 \times 2.1$, which equals 6.3; then simply add the exponential values $10^6$ and $10^3$ ,which is $10^9$; therefore our answer is $6.3 \times 10^9$.  If the exponential value is a negative number, such as $10^{-3}$ , you still add the exponents. So, $10^6$ plus $10^{-3}$ would equal $10^3$. You can see that multiplying with exponents is fairly simple.

When dividing using scientific notation, divide the values between one and ten as you normally would and subtract the exponent values (subtract the exponent value in the denominator from the exponent value in the numerator). For example, divide $6.3 \times 10^{10}$ by $2.1 \times 10^5$; what will be the answer?

In this problem, we divide 6.3 by 2.1, which is 3.0; Next, we subtract the exponent values: $10^{10}$ minus $10^5$ equals $10^5$. Therefore, our final answer is $3.0 \times 10^5$. Suppose the value in the denominator was $2.1 \times 10^{-5}$, what would we do now? The same as before, except when subtracting the exponent values, a minus times a negative is a positive value, and we would add the exponent values. In this case, our new answer would be:

$3.0 \times 10^{15}$                    $^5$.

When adding and subtracting using scientific notation, the exponent values must be to the same exponent value. For example, if you were to add $2.10 \times 10^5$ to $3.20 \times 10^4$, we would have to change one of the values to the same exponent value before we could add the numbers. We could change $2.10 \times 10^5$ to $21.0 \times 10^4$ and then we could add the numbers: $21.0 \times 10^4 + 3.20 \times 10^4$ , which equals $24.2 \times 10^4$. Again, the same rule would apply when subtracting scientific numbers.

# Significant Digits

Significant digits (sometimes called significant figures) are digits that represent measured quantities. Suppose you were measuring the volume of a liquid in a graduated cylinder in the lab, and your graduated cylinder was calibrated to measure to the tenth of a milliliter. You pour in a given amount of liquid into the graduated cylinder and the meniscus (bottom of the curved liquid) rest between two different tenths mark. Utilize the following rules to help determine which numbers are significant and which numbers are not.

## First Rule:

End zeros may or may not be significant if the number has no decimal point. Either express the number in scientific notation or add a decimal point to clarify the number of significant digits. For example: How many significant digits are in the number 80,000? At least one (the digit 8), but the zeros are questionable. To clarify the number of significant digits, we could add a decimal at the end of the number and express it as 80,000., or express it in scientific notation, $8.000 \times 10^4$.  Both of these expressions for eighty thousand represent four significant digits.

## Second Rule:

All digits are significant except zeros at the beginning of a number and possibly zeros at the end of a number. The number 0.208 has three significant digits. The first zero is not significant, it only depicts the size of the number. A better way to express this value would be to write it in scientific notation as $2.08 \times 10^{-1}$. The middle zero is significant because it falls between two whole numbers. The zeros at the end of 2.88000 are significant because they actually indicate that the quantity was measured to the hundred thousandth place. End zeros are significant if the number contains a decimal point as described above.

## Third Rule:

When multiplying or dividing, the final answer can be expressed in only the number of significant digits as the value with the least number of significant digits. For example, consider the following problem: 2.0 x 3.14 x 4.888/ 22.4 x 273.15 = ?

The final answer can only be expressed to two significant digits because that is the smallest number of significant digits in the values given. Also, when one is adding or subtracting numbers, the final answer depends upon the number with the least number of significant digits to the right of the decimal point. For example, what is the correct answer when adding 2.51 plus 3.2? The correct answer is 5.7. The 0.01 value is dropped due to the least number of significant digits to the right of the decimal point being only one.

## Density

Density is one of the physical properties of all matter. By definition, density is mass per volume of a given substance. For example, 10 ml of water has a mass of 10 grams, thus the density of water is 1.0 g/ml.  The units used to express density are g/ml for liquids; g/cm$^3$ for solids and g/L for gases. Note that one gram per milliliter is the same expression as one kilogram per liter. If both the numerator and denominator are multiplied by 1000, the ratio is the same. The units may be different, but the values are the same since the ratio is the same. Since the density of water is 1.0g/ml, it can also be expressed as 1.0 kg/L. When working density problems, be aware that there are three variables: (mass, volume and density). If one knows two of these variables, the third variable can always be calculated. For example, what is the density of a solid element if 193 grams of it displaces 10.0 ml of water? The 10.0 ml of water is equal to 10.0 cm$^3$, since 1.0 ml equals 1.0 cm$^3$. Therefore, dividing the mass of 193 grams by 10.0 cm equals the density, which is 19.3 g/cm$^3$. If someone knows the density and volume of a substance, then the mass can be calculated by rearranging the equation of "density equals mass/volume", such that "mass equals density times volume", or mathematically written as: mass = density x volume. By the same token, if the density and mass are known, then the volume equals mass/density. Density problems involve simple algebra calculations. Here is another problem for you: What is the density of a substance if its mass is 135.5 grams and its volume is 10.00 ml?

## Fahrenheit, Celsius (centigrade), and Kelvin Scales (Thermometers)

When measuring temperature in scientific experiments, one either uses the absolute temperature scale (Kelvin) or the Celsius scale, which is also known as the centigrade scale. The Fahrenheit scale is used in the English system. It is important to be able to convert from one to another. Let us compare the three different temperature scales. To do this, we need a standard to compare the three scales. The freezing point of water and its boiling point at sea level are two good points to use when comparing the three different thermometers.

Let us compare the Fahrenheit and Celsius thermometers first. The freezing temperature of water on the Fahrenheit thermometer is 32 °F, and the freezing temperature of water on the Celsius thermometer is zero degrees. The boiling temperature of water (at sea level) on the Fahrenheit thermometer is 212 °F, and the boiling temperature of water (at sea level) on the Celsius is 100 °C. Note that there are 180 units between 32 °F and 212 °F and 100 units between zero Celsius and 100 °C. Therefore, the ratio of units between the Fahrenheit thermometer and the Celsius thermometer is 180 to 100 or 1.8 to 1.0 unit. So for every 1.0 degree rise in temperature on the Celsius thermometer there is a 1.8 degree rise on the Fahrenheit thermometer.  Also, note that the Fahrenheit thermometer begins at 32 degrees, while the Celsius thermometer begins at zero degrees. To convert from Fahrenheit to Celsius, we subtract 32 degrees from the Fahrenheit temperature and divide by 1.8 to get the Celsius temperature [( °F -32)/1.8 = ° C].   Problem: Let's convert body temperature of 98.6 °F to Celsius: (98.6 – 32)/1.8 = 37 °C. To convert from Celsius to Fahrenheit, multiply the Celsius temperature by 1.8 then add 32 degrees. [( °C x 1.8) + 32° = °F]

Another problem: If room temperature is 68 °F, what is this temperature in Celsius? _____

What do we do when we want to convert from Celsius to Kelvin? That is a simple conversion. Take the Celsius temperature and add 273, and you have the Kelvin temperature. Twenty-five degrees Celsius is 298 Kelvin. (Note we do not use the term degrees when working with Kelvin).

For practice, make conversions between the three different thermometers by using temperatures of the melting point and boiling point of various elements given on the Periodic Table.

## Energy and Measurement

A few definitions are necessary when working with energy problems. Energy in the form of heat can be measured in either calories or joules. *One calorie is the amount of heat required to raise the temperature of 1.0 gram of $H_2O$ one degree Celsius*; and one calorie is equal to 4.184 joules.

In a similar fashion, one joule is equal to 0.239 calories. Sometimes, these calories are referred to as small calories. They are not the same as Food calories. It takes one thousand of these small calories to make one Food calorie. Therefore, one Food calorie is equal to one kilocalorie (kcal).

The relationship between calories and joules can be expressed mathematically in the formulas:

calorie = 4.184 joules and 1.0 joule = 0.239 calories   (Additional Factor Conversions)

Let's consider a problem that involves converting from calories to joules and vice versa. Suppose you have 1.507 x 10⁶ calories on hand, how many joules would this equal? Remember to round off the answer to the necessary number of significant digits. One could multiply the 1.507 x 10⁶ calories by 4.184 joules/calorie and the calorie units would cancel, leaving an answer in joules. This would equal 6.305 x 10⁶ joules.

Next, convert 9.44 kilojoules into calories. Did you get 2.26 x 10³ calories?

**Specific Heat** is the amount of heat required to raise the temperature of a one gram of substance one degree Celsius. All substances have specific heats. For example, the specific heat of water is 1.0 calorie/ gram · °C or 4.184 joules/ gram · °C. Suppose you wanted to calculate how many calories it would take to raise the temperature of 100.0 grams of water from 25.0 °C to 75.0 °C. What would you do? The amount of heat required is equal to the mass of substance in grams times its specific heat times the change in temperature. Use Celsius units in your calculations.

Amount of Heat =  mass (grams)  x  specific heat  x  temperature change   or to simplify:

Q = mass  x  sp. ht.  x  $\Delta T$   where Q = amount of heat  and  $\Delta T$ = temperature change

To solve our problem, we have Q = 100.0 g x 1.0 cal/g · °C x 50.0 °C  Note that the temperature change is 50 °C in this problem. Our answer is: 5.00 x 10³ calories of 5.00 kcal.

## Dimensions

Dimensions are fundamental quantities such as length, mass and time, whereas SI units are meters, kilograms, and seconds. EXAMPLE: area of a rectangle: dimension- Length squared, or $L^2$ and SI unit is meter squared or $m^2$.

Dimensions could be used to check the validity of a relationship (i.e. a formula).

If the dimensions on one side of a formula do not equal the other side, then the formula is incorrect. If the dimensions on one side of a formula are equal to the other side then the formula is correct unless it has some other constants in the formula. In such cases, just by comparing dimensions, we cannot say that the formula is correct or not.

Examples:

Mass = M

Length = L

Time = T

Electric Charge = Q

Volume = $L^3$

Acceleration (velocity/time) = $L/T^2$

Density (mass/volume) = $M/L^3$

Force (mass × acceleration) = $M \cdot L/T^2$

Charge (current × time) = $I \cdot T$

# Additional applications of dimensions to other measurements:

| | |
|---|---|
| Pressure (force/area) | $M \cdot L^{-1} \cdot T^{-2}$ |
| (Volume)² | $L^6$ |
| Electric field (force/charge) | $M \cdot I^{-1} \cdot T^{-3}$ |
| Work (in 1-D, force × distance) | $M \cdot L^2/T^2$ |
| Energy (e.g., gravitational potential energy = mgh = mass × gravitational acceleration × height) | $M \cdot L^2/T^2$ |
| Square root of area | L |

## The resolving of vectors is discussed under chapter 3.

For practice questions and answers on topics covered here, please visit:

http://student.ccbcmd.edu/~cyau1/121PractProbOnUnitConversionSp2005.pdf

http://fountain.cnx.rice.edu:8280/content/m15037/latest/

# PowerPoint Link

Please refer to the end of the module lecture links.

# Discussion Question

Answer the following questions with references. Please remember to follow the standard APA referencing style.

For APA standards of references, please visit:

http://owl.english.purdue.edu/owl/resource/560/01/

Also, respond in detail to one other post by fellow students.

1.1 Visit the website: http://www.cnn.com/TECH/space/9909/30/mars.metric.02/ and read about "Metric mishap caused loss of NASA orbiter." Explain the importance of a universal standard in measurements. Elaborate your answer in terms of metric system and conversions.

1.2 Describe the importance of a universal standard measuring system (SI Units) in scientific work with examples from real-life applications.

# Laboratory Activity and the link

Go to the link below and run the simulation:

http://phet.colorado.edu/simulations/sims.php?sim=Motion_in_2D

Go to: http://phet.colorado.edu/teacher_ideas/view-contribution.php?contribution_id=627

Open Phet 2D Motion Activity.doc

Answer all the questions.

_____ **(not included with this book)**

# M o d u l e 1

## C H A P T E R - 2

# Motion in one Dimension

## Objectives

At the end of this lesson, you should be able to:

1.  Define scalars, vectors, distance, displacement, speed, velocity, and acceleration;

2.  Apply kinematic equations;

3.  Analyze various position, velocity, and acceleration graphs;

4.  Calculate freefall speed, distance, etc.

## Lecture Notes

### Distance and Displacement

When a particle changes its position, it is said to have undergone a certain distance and if it is expressed with reference to a direction then this distance is known as displacement.

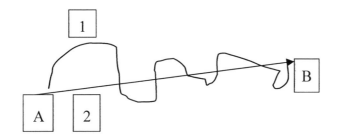

**Figure 2.1**

When a particle moves from A to B, either by using path 1 or 2, it has undergone a displacement shown by the straight line. The displacement for both pathways is the same. It is given as the magnitude of the direct distance from A to B, expressed with a direction.

This is a vector, which has both a magnitude and a direction.

The distance of path 1 is longer than the distance of path 2. These are known as scalars, which only have a magnitude (no direction). Both a scalar and a vector have units. In this case, it is meters.

- Some examples of scalars are speed, time, mass, density, temperature, and energy.
- Scalar quantities can be added, subtracted, multiplied, or divided by the ordinary rules.
- Some examples of vectors are force, velocity, acceleration, weight, electric field strength, and magnetic strength.
- There are rules to follow when operating with vector quantities.

Kinematics refers to the study of motion of natural bodies. The bodies that we see and deal with in real life are three-dimensional objects and essentially not a point object.

When a body moves *with* rotation (rolling while moving), the path trajectories of different parts of the bodies are different; on the other hand, when the body moves *without* rotation (slipping/ sliding), the path trajectories of the different parts of the bodies are parallel to each other.

In the second case, the motion of all points within the body is equivalent as far as translational motion of the body is concerned and hence, such bodies may be said to move like a point object. It is, therefore, possible to treat the body under consideration to be equivalent to a point so long as rotation is not involved.

For this reason, the study of kinematics consists of studies of:

1. Translational kinematics
2. Rotational kinematics

A motion can be pure translational or pure rotational or a combination of the two types of motion.

The translational motion allows us to treat a real time body as a point object. Hence, we freely refer to bodies, object, and particles in the same sense that all of them are point entities, whose position can be represented by a single set of coordinates. We should keep this in mind while studying translational motion of a body and treating the same as point.

## Distance – time plot

A "distance – time" plot is a simple plot of two scalar quantities along two axes. However, the nature of distance imposes certain restrictions, which characterize a "distance – time" plot.

The nature of a "distance – time" plot, with reference to its characteristics, is summarized here:

1. Distance is a positive scalar quantity. As such, "the distance – time" plot is a curve in the first quadrant of the two dimensional plot.
2. As distance keeps increasing during a motion, the slope of the curve is always positive.

3.  When the object undergoing motion stops, then the plot becomes a straight line parallel to the time axis, so that distance is constant as shown in the figure here:

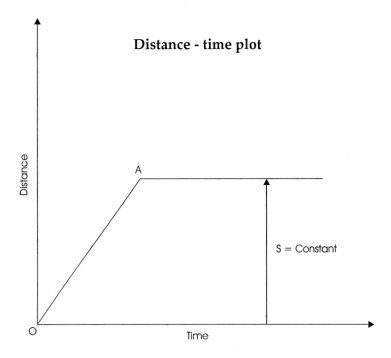

**Figure 2.2**

One important implication of the positive slope of the "distance – time" plot is that the curve never drops below a level at any moment of time. Besides, it must be noted that the "distance - time" plot is handy in determining "instantaneous speed", but we choose to conclude the discussion of "distance - time" plot as these aspects are separately covered in subsequent module.

# Example 1: Distance – time plot

*Question* : A ball falling from an height 'h' strikes the ground. The distance covered during the fall at the end of each second is shown in the figure for the first 5 seconds. Draw a "distance – time" plot for the motion during this period. Also, discuss the nature of the curve.

## Motion of a falling ball

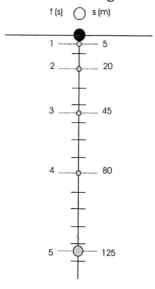

**Figure 2.3**

*Solution* : We have experienced that a free falling object falls with increasing speed under the influence of gravity. The distance covered in successive time intervals increases with time. The magnitudes of distance covered in successive seconds given in the plot illustrate this point. In the plot between distance and time as shown, the origin of the reference (coordinate system) is chosen to coincide with initial point of the motion.

### Distance – time plot

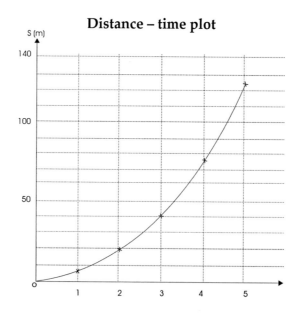

**Figure 2.4**

From the plot, it is clear that the ball covers more distance as it nears the ground. The "distance- time" curve during fall is, thus, flatter near the start point and steeper near the earth's surface. Can you guess the nature of the plot when a ball is thrown up, against gravity?

# Exercise 1

A ball falling from a height 'h' strikes a hard horizontal surface with increasing speed. On each rebound, the height reached by the ball is half of the height it fell from. Draw a "distance – time" plot for the motion covering two consecutive strikes, emphasizing the nature of curve (ignore actual calculation). Also, determine the total distance covered during the motion.

### Solution

Here we first estimate the manner in which distance is covered under gravity as the ball falls or rises.

The "distance- time" curve during fall is flatter near the start point and steeper near the earth's surface. On the other hand, we can estimate that the "distance- time" curve, during rise, is steeper near the earth's surface (covers more distance due to greater speed) and flatter as it reaches the maximum height, when the speed of the ball becomes zero.

The "distance – time" plot of the motion of the ball, showing the nature of curve during motion, is:

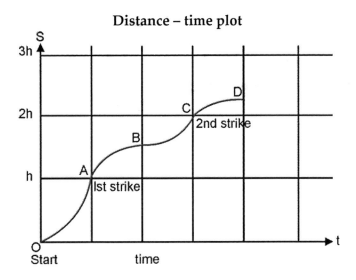

**Figure 2.5**

The origin of plot (O) coincides with the initial position of the ball (t = 0). Before striking the surface for the first time (A), it travels a distance of 'h'. On rebound, it rises to a height of 'h/2' (B on the plot). Total distance is 'h + h/2 = 3h/2'. Again falling from a height of 'h/2', it strikes the surface, covering a distance of 'h/2'. The total distance from the start to the second strike (C on the plot) is '3h/2 + h/2 = 2h'.

The coordinate system enables us to specify a point in its defined volumetric space. We must recognize that a point is a concept without dimensions; whereas the objects or bodies under motion themselves are not points. The real bodies, however, approximates a point in translational motion, when paths followed by the particles, composing the body are parallel to each other (See Figure). As we are concerned with the geometry of the path of motion in kinematics, it is, therefore, reasonable to treat real bodies as "point like" mass for description of translational motion.

Translational motion

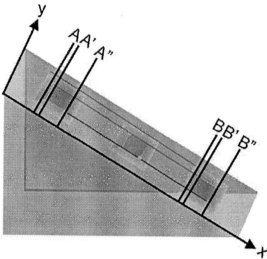

**Figure 2.6: Particles follow parallel paths**

We conceptualize a particle in order to facilitate the geometric description of motion. A particle is considered to be dimensionless, but having a mass. This hypothetical construct provides the basis for the logical correspondence of a point, with the position occupied by a particle.

Without any loss of purpose, we can designate motion to begin at A or A′ or A″, corresponding to final positions B or B′ or B″ respectively, as shown in the figure above.

For the reasons outlined above, we shall freely use the terms "body" or "object" or "particle" in the same way as far as the description of translational motion is concerned. Here, pure translation conveys the meaning that the object is under motion without rotation, like sliding of a block on a smooth inclined plane.

# Position

## Definition 1: Position

The position of a particle is a point in the defined volumetric space of the coordinate system.

**Position of a point object**

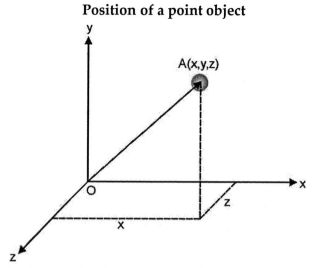

**Figure 2.7: Position of a point is specified by three coordinate values**

The position of a point-like object, in three dimensional coordinate space, is defined by three values of coordinates i.e. x, y and z in the Cartesian coordinate system as shown in the figure above.

It is evident that the relative position of a point with respect to a fixed point, such as the origin of the system "O", has directional property. The position of the object, for example, can lie either to the left or to the right of the origin, or at a certain angle from the positive x - direction. As such the position of an object is associated with directional attribute with respect to a frame of reference (coordinate system).

# Example 1: Coordinates

*Problem* : The length of the second's hand on a round wall clock is 'r' meters. Specify the coordinates of the tip of the second's hand corresponding to the markings 3,6, 9 and 12 (Consider the center of the clock as the origin of the coordinate system.).

**Coordinates of the tip of the second's hand**

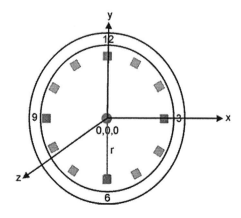

**Figure 2.8: The origin coincides with the center**

*Solution* : The coordinates of the tip of the second's hand is given by the coordinates :

3 :  r, 0, 0
6 :  0, -r, 0
9 :  -r, 0, 0
12 :  0, r, 0

# Plotting motion

The position of a point in the volumetric space is a three-dimensional description. A plot showing positions of an object during a motion is an actual description of the motion as far as the curve shows the path of the motion and its length gives the distance covered. A typical three-dimensional motion is depicted as in the figure below:

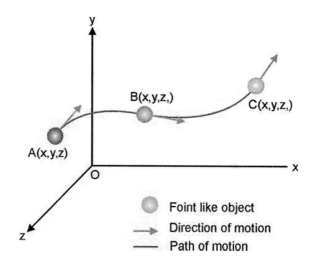

**Figure 2.9: The plot is the path followed by the object during motion**

In the figure, the point like object is deliberately shown not as a point, but with finite dimensions. This has been done in order to emphasize that an object of finite dimensions can be treated as point when the motion is purely translational.

The three-dimensional description of positions of an object during motion is reduced to be two or one dimensional description for the planar and linear motions respectively. In two or one dimensional motion, the remaining coordinates are constant. In all cases, however, the plot of the positions is meaningful in following two respects :

• The length of the curve (i.e. plot) is equal to the distance.

• A tangent in forward direction at a point on the curve gives the direction of motion at that point

# Description of motion

Position is the basic element used to describe motion. Scalar properties of motion, like distance and speed, are expressed in terms of position as a function of time. As the time passes, the positions of the motion follow a path, known as the trajectory of the motion. It must be emphasized here that the path of motion (trajectory) is unique to a frame of reference, and so is the description of the motion.

To illustrate the point, let us consider that a person is traveling on a train, which is moving with the velocity $v$ along a straight track. At a particular moment, the person releases a small pebble. The pebble drops to the ground along the vertical direction as seen by the person.

An observer on the ground, however, sees the same incident, as if the pebble followed a parabolic path (See Figure blow). It emerges then that the path or the trajectory of the motion is also a relative attribute, like other attributes of the motion (speed and velocity). The coordinate system of the passenger in the train is moving with the velocity of train ($v$) with respect to the earth and the path of the pebble is a straight line. For the person on the ground, however, the coordinate system is stationary with respect to earth. In this frame, the pebble has a horizontal velocity, which results in a parabolic trajectory.

**Trajectory as seen by the person on ground**

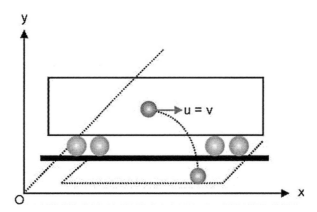

**Figure 2.10: Trajectory is a parabolic curve.**

Without overemphasizing, we must acknowledge that the concept of path or trajectory is essentially specific to the frame of reference or the coordinate system attached to it. Interestingly, we must be aware that this particular observation happens to be the starting point for the development of the special theory of relativity by Einstein (see his original transcript on the subject of relativity).

## *Position – time plot*

The position in three-dimensional motion involves specification in terms of three coordinates. This requirement poses a serious problem when we want to investigate positions of the object with respect to time. In order to draw such a graph, we would need three axes for describing position and one axis for plotting time. This means that a position – time" plot for a three-dimensional motion would need four (4) axes !

A two-dimensional "position – time" plot, however, is a possibility, but its drawing is highly complicated for representation on a two-dimensional paper or screen. A simple example consisting of a linear motion in the x-y plane is plotted against time on z – axis (See Figure).

**Two dimensional position – time plot**

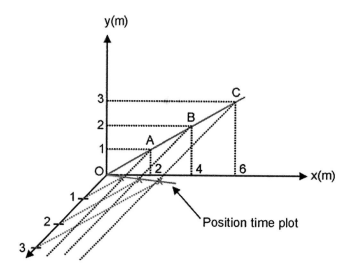

**Figure 2.11:**

In fact, it is only the one-dimensional motion whose "position – time" plot can be plotted conveniently on a plane. In one-dimensional motion, the point object can be either to the left or to the right of the origin in the direction of the reference line. Thus, drawing position against time is a straightforward exercise as it involves plotting positions with the appropriate sign.

# Example 2: Coordinates

*Problem* : A ball moves along a straight line from O to A to B to C to O along the x-axis as shown in the figure. The ball covers each of the distances of 5 m in one second. Plot the "position – time" graph.

**Motion along a straight line**

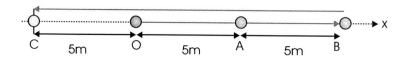

**Figure 2.12:**

*Solution* : The coordinates of the ball are 0,5,10, -5 and 0 at points O, A, B, C and O (on return) respectively. The "position – time" plot of the motion is as given below :

**Position – time plot in one dimension**

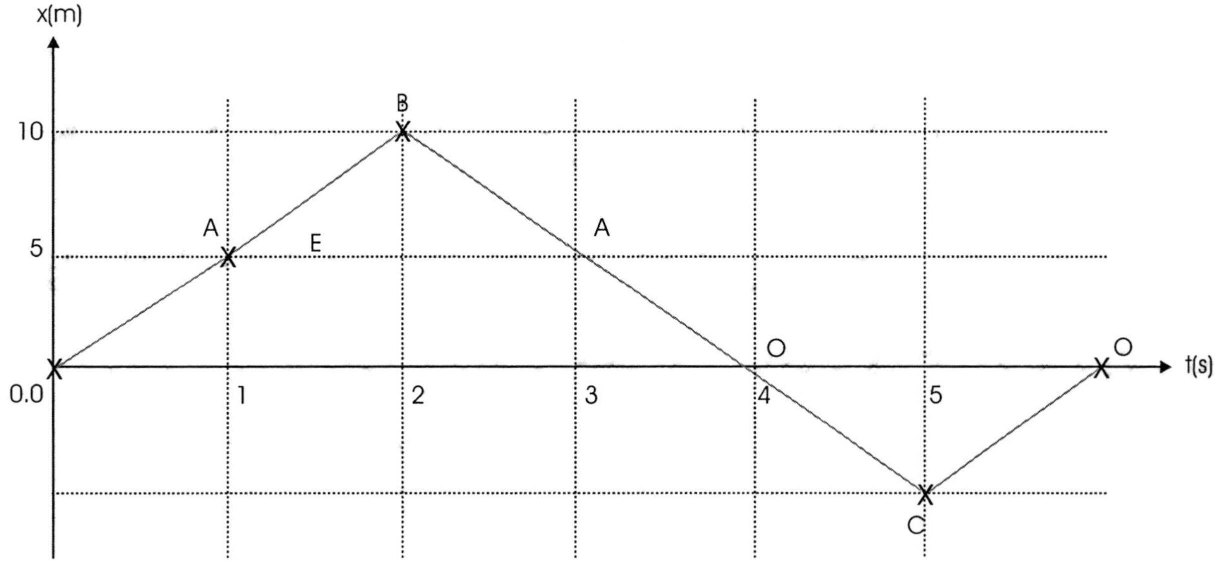

**Figure 2.13:**

# Speed and velocity

Speed is a scalar quantity. Velocity is a vector. Speed is given as the distance of motion per unit time. The velocity is the displacement (direction is noted) per unit time.

Both have the same magnitude and same units (meters per second, ms⁻¹) but only the velocity has a direction specified.

## Average Speed and average velocity

Average speed is the total distance over the total period of time:

$$\text{Average Speed} = \frac{\text{Distance Traveled}}{\text{Time of Travel}}$$

Average velocity is the total displacement over the total period of time:

$$\overline{V} = \frac{\Delta x}{\Delta t}$$

$$\text{Average Velocity} = \frac{\Delta \text{position}}{\text{time}} = \frac{\text{displacement}}{\text{time}}$$

# Instantaneous speed and instantaneous velocity

Instantaneous speed is the speed at any given instant in time. (The speedometer in a car indicates the instantaneous speed)

Instantaneous velocity is the velocity at any given instant time. This is given by,

$$v = \lim_{\Delta t \to 0} \frac{\Delta x}{\Delta t}$$

which becomes,

$$v = \frac{dx}{dt}$$

# Constant speed and constant velocity

If there is no variation in the speed (or velocity) during a specific movement, then that object is said to be at constant speed (or velocity).

# Velocity and acceleration

The rate of change of velocity is known as the acceleration. The change of velocity can be an increase in velocity, a decrease in velocity or a change in direction.

$$\text{Average acceleration} = \frac{\Delta \text{velocity}}{\text{time}} = \frac{v_f - v_i}{t}$$

$V_f$ = Final velocity

$V_i$ = Initial velocity

t = Time

Constant acceleration is when the change in velocity per second is constant.

Instantaneous acceleration is given by,

$$a = \lim_{\Delta t \to 0} \frac{\Delta v}{\Delta t}$$

$$a = \frac{dv}{dt}$$

The relationship between the displacement and the acceleration is the second derivative of the position:

$$a\,(t) = \frac{dv}{dt} = \frac{d\left(\frac{dx}{dt}\right)}{dt} = \frac{d^2x}{dt^2}$$

It is important to note that the velocity can be directed in a positive direction or a negative direction (The decision of a positive direction or negative direction is a relative factor. Normally the direction of movement is considered to be positive. e.g.  An object falling freely downward is at a positive direction and the opposite is a negative direction. )

* If velocity is in a positive direction and if the velocity is increasing with time, then the acceleration is said to be positive. In this case, if the velocity is reducing with time, then the acceleration is said to be negative.

* If velocity is in a negative direction and if the velocity is increasing with time then the acceleration is said to be negative. In this case, if the velocity is reducing with time, then the acceleration is said to be positive.

# Graphs, Derivations and calculations

The speed, velocity, and acceleration can be expressed in graphical form.

| Positive Constant Velocity | Positive Changing Velocity |
|---|---|
| 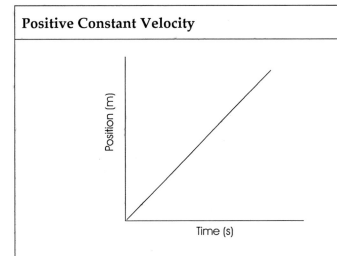 | 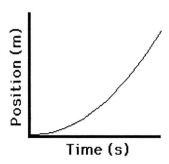 |
| According to this graph the acceleration is zero as there is no change in velocity. The slope of the graph indicates the velocity changes. Here the change is constant. If two such graphs are compared then the graph with higher slope will be considered to have a higher velocity. A negative velocity on a graph indicates a downward slope. | According to this graph, the velocity is changing, hence there is an acceleration. The indication that the graph is curved upward indicates an increase in velocity. A negative changing velocity can be shown if the above graph is inversed. |

**Diagram 214**

The area under a velocity vs. time graph gives the displacement.

The area under an acceleration vs. time graph gives the velocity.


Position ----------------→ Velocity ------------------→ Acceleration

   Differentiate                Differentiate


Position <--------------- Velocity <-----------------Acceleration

   Integrate                    Integrate

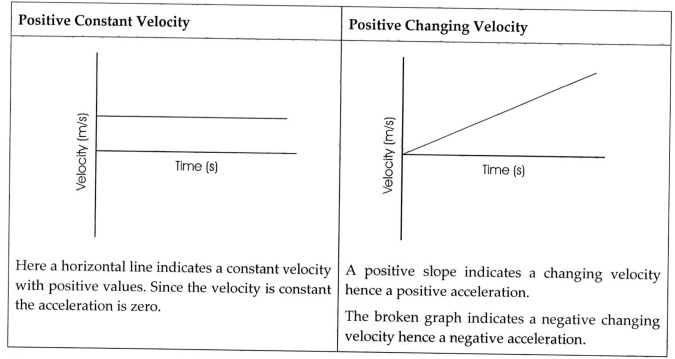

| Positive Constant Velocity | Positive Changing Velocity |
|---|---|
| Here a horizontal line indicates a constant velocity with positive values. Since the velocity is constant the acceleration is zero. | A positive slope indicates a changing velocity hence a positive acceleration. The broken graph indicates a negative changing velocity hence a negative acceleration. |

**Diagram 2.15**

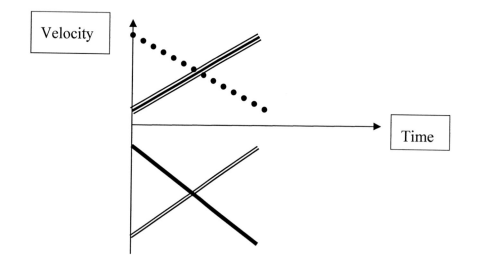

**Diagram 2.16**

Consider each graph in diagram 2.16

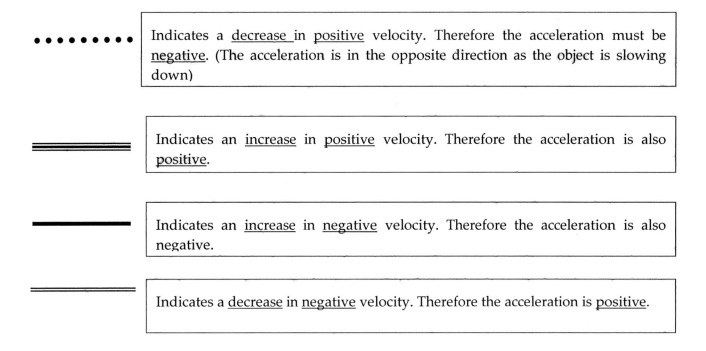

Indicates a <u>decrease</u> in <u>positive</u> velocity. Therefore the acceleration must be <u>negative</u>. (The acceleration is in the opposite direction as the object is slowing down)

Indicates an <u>increase</u> in <u>positive</u> velocity. Therefore the acceleration is also <u>positive</u>.

Indicates an <u>increase</u> in <u>negative</u> velocity. Therefore the acceleration is also negative.

Indicates a <u>decrease</u> in <u>negative</u> velocity. Therefore the acceleration is <u>positive</u>.

## Free Fall

Free fall relates to any object falling down towards the Earth. In this situation, the resistance due to the upward thrust of air is ignored. Therefore, the only force acting upon it is the gravity. The downward acceleration is 9.8 m s⁻². This is known as the **acceleration of gravity** and is symbolized as **g**.

The conventional signs for vectors such as velocity and acceleration are upward positive sign and downward negative sign,, and right hand direction positive and left and direction negative.

For a free falling object, the downward direction can be taken as negative. Therefore, a graph of position vs. time and velocity vs. time would appear as follow:

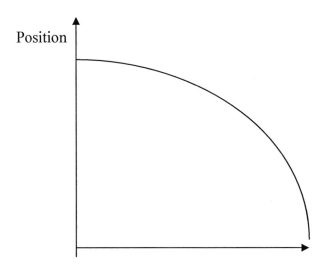

**Diagram 2.17**

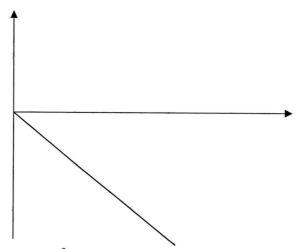

Slope indicates an acceleration of $-10 \text{ m s}^{-2}$

**Diagram 2.18**

In order to calculate the velocity of a free falling object after a certain period of time, one can use the velocity-acceleration equation.

acceleration = velocity / time

∴ **velocity = acceleration X time**                                                              2.1

Using this equation for a free falling equation, at any given time the velocity can be found. Acceleration of gravity is constant which is 9.8 m s⁻². 

∴ **velocity = 9.8 X time   ms⁻¹**                                                               2.2

Similarly, the downward distance travelled of a falling object at a given time can be calculated by deriving the following equation:

Initial velocity of an object falling downward from rest = 0

Velocity at time t   = v

The average velocity   =   ( 0 + v ) / 2 = 0.5 v

If  velocity = distance / time

Then distance travelled = 0.5v  X  t

From equation 4.2 ....


v = 9.8 X time

∴ distance travelled = 0.5 X  9.8 X t X t = 0.5 X 9.8 X t²


9.8 is rounded off to 10, for convenience.

∴ distance travelled (s) = 0.5 X 10 X t²                                                        2.3

or

Distance travelled (s) = 2 t²                                                                                    2.4

Irrespective of the size, all objects free fall under the same acceleration of gravity. This means, in the absence of all the forces except for acceleration of gravity, all objects fall at the same acceleration. Therefore, two items with varying masses dropped at the same time will reach the ground at the same time. In this case, we ignore the air resistance. The shape and the size (hence the area covered) of the falling object has an impact when the air resistance is considered.

## Kinematics Equations

Notations:

s = distance (x is also used for distance)

v = velocity

f = final

i = initial

a = acceleration

t = time

The attempt here is to derive several equations that will enable determination of any of the above-notated variables.

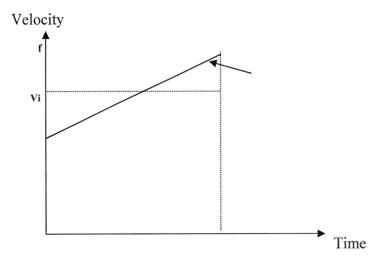

**Diagram 2.19**

The above diagram shows a graph of a moving object indicating its velocity against time.

In a graph, the area under the graph is the total distance travelled, which is the total of A and B.

$V_i$ is the initial velocity and $V_f$ is the final velocity and t is the time travelled.

Therefore, the area under the curve can be expressed as:

**For section B = $v_i$ X t**

For section A = ($v_f - v_i$ ) X t X  0.5

The total distance travelled (s)…

$$s = A + B = t\,(v_i) \ + 0.5t(v_f - v_i)$$

**s = t ($v_i$)   + 0.5t($v_f - v_i$)**                                        2.5

∴ s = $tv_i$ + 0.5$tv_f$ − 0.5$tv_i$  = 0.5$tv_i$ + 0.5 $tv_f$ = 0.5t ($v_i$ + $v_f$)

**__s__  = 0.5t ($v_i$ + $v_f$)**                                        2.6

From the slope of the graph, we can calculate the acceleration. Therefore;

Slope = acceleration (a) = ($v_f - v_i$) / t

**at = $v_f - v_i$,**                                        2.7

by rearranging….

**$v_f$ = $v_i$ + at**                                        2.8

By substituting 2.8 to 2.5,  we obtain….

$$s = t\,(v_i) \ + 0.5t\,(at)$$

∴    **s = t ($v_i$)   + 0.5at²**                                        2.9

Equation 2.7 can be rearranged as ….

t = ($v_f - v_i$) / a , and by substituting this to the equation 2.6 we get….

S= 1/2$a$  ($v_f - v_i$) ($v_i$ + $v_f$)        (1/2a is the same as 0.5/a)

by rearranging the above equation we obtain…

$$2as \ = v_f{}^2 - v_i{}^2$$

By rearranging the above equation, we obtain…

**$v_f{}^2$ = $v_i{}^2$ + 2as**                                        2.10

The above underlined four equations are the most important ones in kinematics.

The instances when initial velocity is zero are, when you are given the following terms:

Start from rest

Object fell

Object was dropped

For questions and answers on kinematics, please visit the website:

http://fountain.cnx.rice.edu:8280/content/m13580/latest/ and then go to subtopics within that page.

# PowerPoint Link:

Please refer to the end of the module lecture links.

# Discussion Question

Answer the following questions with references. Please remember to follow the standard APA referencing style.

For APA standards of references, please visit: http://owl.english.purdue.edu/owl/resource/560/01/

Also, respond in detail to one other post by fellow students.

2.1 Explain the changes in the velocity, acceleration, and net force of a man with a parachute jumping off a plane. You need to describe the changes with valid scientific principles.

2.2 Explain the scientific theory behind free fall of an acorn and pumpkin to an atheist who is also a non-scientist. You may visit the website: http://digital.library.upenn.edu/women//finch/1713/mp-atheist.html to read more on the poem, "The Atheist and the Acorn."

# Laboratory Activity and the link

Go to the link below and run the simulation:

http://phet.colorado.edu/simulations/sims.php?sim=The_Moving_Man

Go to: http://phet.colorado.edu/teacher_ideas/view-contribution.php?contribution_id=31 and open mm2.doc

Answer all the questions.


Now take the chapter 2 test (not included with this book)

# Module 1

## CHAPTER-3

# Vectors and Two Dimensional Motions

## Objectives

At the end of this lesson, you should be able to:

1.    Resolve vectors;

2.    Calculate vector components;

Carry out two-dimensional analysis and solve problems.

## Lecture Notes

A coordinate system is a system of measurements involving distance and direction with respect to rigid bodies. Structurally, it is comprised of coordinates and a reference point, usually the origin of the coordinate system. The coordinates primarily serve the purpose of reference for the direction of motion, while the origin serves the purpose of reference for the magnitude of motion.

Measurements of magnitude and direction allow us to locate the position of a point in terms of measurable quantities like linear distances or angles, or their combinations. With these measurements, it is possible to locate a point in the spatial extent of the coordinate system. The point may lie anywhere in the spatial (volumetric) extent defined by the rectangular axes as shown in figure 3.1 (note : the point in the figure is shown as a small sphere for visual emphasis only).

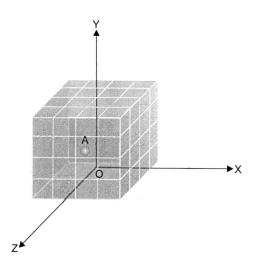

**Figure 3.1**

A distance in the coordinate system is measured with a standard rigid linear length like that of a "meter" or a "foot". A distance of 5 meters, for example, is 5 times the length of the standard length of a meter. On the other hand, an angle is defined as a ratio of lengths and is dimensional-less. Hence, the measurement of direction is indirectly equivalent to the measurement of distances only.

The coordinate system represents the system of rigid bodies, like earth, which is embodied by an observer making measurements. Since measurements are implemented by the observer, they (the measurements in the coordinate system) represent distance and direction as seen by the observer. It is, therefore, clearly implied that measurements in the coordinates system are specific to the state of motion of the coordinate system.

## Coordinate system types

Coordinate system types determine the position of a point with measurements of distance or angle, or a combination of them. A spatial point requires three measurements in each of these coordinate types. It must be noted that the descriptions of a point in any of these systems are equivalent. Different coordinate types are merely for the convenience of appropriateness for a given situation. Three major coordinate systems used in the study of physics are:

- Rectangular (Cartesian)
- Spherical
- Cylindrical

The rectangular (Cartesian) coordinate system is the most convenient, as it is easy to visualize and associate with our perception of motion in daily life. Spherical and cylindrical systems are specifically designed to describe motions, which follow spherical or cylindrical curvatures.

# Rectangular (Cartesian) coordinate system

The measurements of distances along three mutually perpendicular directions, designated as x,y and z, completely define a point A (x,y,z).

## A point in a rectangular coordinate system

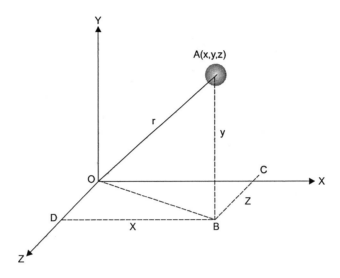

**Figure 3.2: A point is specified with three coordinate values**

From geometric consideration of triangle OAB,

$$r=\sqrt{OB^2+AB^2}$$

From geometric consideration of triangle OBD,

$$OB^2=\sqrt{BD^2+OD^2}$$

Combining above two relations, we have :

$$\Rightarrow r=\sqrt{BD^2+OD^2+AB^2}$$

$$\Rightarrow r=\sqrt{x^2+y^2+z^2}$$

# Spherical coordinate system

A three-dimensional point "A" in a spherical coordinate system is considered to be located on a sphere of a radius "r". The point lies on a particular cross section (or plane) containing the origin of the coordinate system. This cross section makes an angle "θ" from the "zx" plane (also known as the longitude angle). Once the plane is identified, the angle, $\phi$, that the line joining origin O to the point A, makes with "z" axis, uniquely defines the point A (r, θ, $\phi$).

## Spherical coordinate system

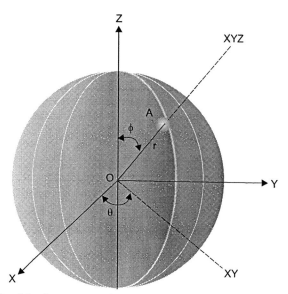

**Figure 3.3: A point is specified with three coordinate values**

It must be realized here that we need to designate three values r, θ and φ to uniquely define the point A. If we do not specify θ, the point could then lie in any of the infinite numbers of possible cross sections through the sphere.

# Cylindrical coordinate system

A three-dimensional point "A" in a cylindrical coordinate system is considered to be located on a cylinder of a radius "r". The point lies on a particular cross section (or plane) containing the origin of the coordinate system. This cross section makes an angle "θ" from the "zx" plane. Once the plane is identified, the height, z, parallel to the vertical axis "z" uniquely defines the point A(r, θ, z)

## Cylindrical coordinate system

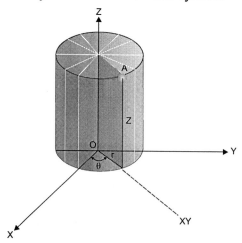

**Figure 3.4: A point is specified with three coordinate values**

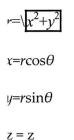

$x = r\cos\theta$

$y = r\sin\theta$

$z = z$

$\text{Tan}\theta = y/x$

Vectors operate with other scalar or vector quantities in a particular manner. Unlike scalar algebraic operations, vector operation draws on graphical representation to incorporate directional aspect.

Vector addition is, however, limited to vectors only. We cannot add a vector (a directional quantity) to a scalar (a non-directional quantity). Further, vector addition is dealt with in three conceptually equivalent ways:

1.  graphical methods

2.  analytical methods

3.  algebraic methods

## Addition of Vectors

Vector diagrams represent vector quantities. This means they are shown by an arrow drawn to a scale (to indicate the magnitude) in a specific direction.  The vector direction can be expressed as a counterclockwise angle of rotation about its tail end from due east.

When vectors are on the same line (the same horizontal and vertical lines), then they may be added according to general algebraic additions.

e.g. :  +5 $\rightarrow$ +  +5 $\rightarrow$  = +10 ———▶

[+5] ◀——— [-10]  ◀——— [-5]

When the vectors are not on the same horizontal or vertical directions, then any of the following two methods may be used:

*   The scaled vector diagram method (the geometric method; the arrows are drawn according to a scale from head to tail)

*   The Pythagorean Theorem and trigonometric method (the analytical method)

## The scaled vector diagram method (geometric method)

e.g.:

Vector A = 30m/s at 45 degrees angle

Vector B = 20 m/s at 90 degrees angle

Add A and B and find the resultant vector.

The first step is to draw the vectors according to a scale.

If you select 10 m as 1 cm, then draw vector A as 3 cm long and at a 45 degree angle (see the drawings below; not drawn according to a scale.) The angle of 45 degrees is measured from the horizontal broken line to the arrow.

Vector A

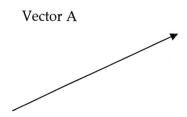

The second step is to draw vector B according to the same scale. The beginning of vector B will be from the end of vector A. This means, you need to draw vector B from the arrowhead of A, according to the scale and the angle.

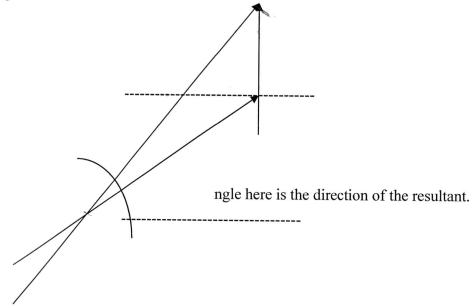

ngle here is the direction of the resultant.

The length of the resultant arrow, drawn from the beginning of vector A to the end of vector B, will be the final vector when you convert the scale back to a value in m/s.

## Resolution of Vectors

Resolution of vectors means the separation of vectors into their components in x,y (and z for three dimensional situations) directions.

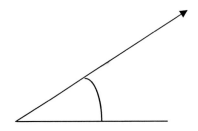

e.g

**Figure 3.5**

If the angle between the horizontal and the vector (arrow) is θ, then by resolving the vector into its components we get for the x axis, Vector X cos θ and for the y axis, Vector X sin θ.

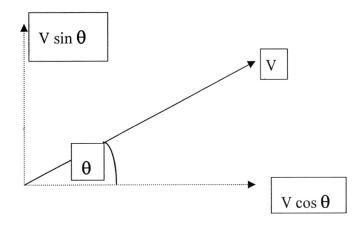

**Figure 3.6**

# The trigonometric method (the analytical method)

In this method, if you take the same example as above, you resolve to its x and y components. Then you add all the x components separately and all the y components separately.

So, by using the same example above, the x component for vector A is 30cos 45 and the x component for vector B is 20 cos 90.

Therefore,

for vector A, it is = 30 cos 45 = 21.21

for vector B, it is = 20 cos 90 = 0

Now add all of the y components for vector A and vector B.

Therefore,

for vector A, it is = 30 sin 45 = 21.21

for vector B, it is = 20 sin 90 = 20

Now add all x values.

The total for x = 21.21

Now add all y values.

The total for y is = 41.21

Now use the Pythagorean Theorem to solve for the resultant:

$$Z^2 = X^2 + Y^2$$

Therefore,

$$Z^2 = 449.86 + 1698.26 = 2148.12$$

$$Z = 46.35 \text{ m/s}$$

This gives the magnitude of the resultant.

Now to get the direction of the resultant (angle) vector, you need to divide the y value by the x value, which gives the tangent of the angle:

$$\text{Tan } \theta = 41.21 \div 21.21 = 1.94$$

Therefore,

$$\theta = 62.73 \text{ degrees.}$$

# Multiplication of Vectors

There are three types of operations:

# Multiplication of a Vector and a Scalar

In this case, the resultant is a vector. The multiplication of a vector (a) by a scalar (k) leads to a vector product that is 'k' times the magnitude of 'a'. If 'k' has a positive value then the product will have the same direction as the original vector. If 'k' has a negative value then the product will have an opposite direction to the original vector.

# Multiplication of two Vectors to get a Scalar product (dot product)

Assume two vectors |a| and |b|. The product is given by $a \cdot b = ab \cos \theta$. The resultant here is a scalar product.

Some of the examples of this application are in finding the work done by force times distance and the potential energy calculations by mgh.

(Vectors are normally denoted as |a| or $\bar{a}$. The vectors are also expressed as **i** for the x axis, **j** for the y axis and **k** for the z axis)

(In the following case, vector |a| can be resolved to its two dimensional components by |a | = |i|
a$_x$   | | y)

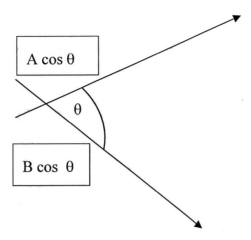

**Figure 3.7**

# Multiplication of two Vectors to get a Vector product (cross product)

Assume two vectors a and b. The resultant vector is given by **a x b = a b sin θ.**

Some examples of this application are in finding torque, angular momentum, or the force of an electric charge in a magnetic field (more details below).

**Direction of the movement**

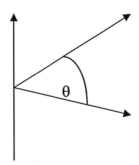

**Figure 3.8**

# Vector addition : Algebraic method

Graphical method is meticulous and tedious as it involves the drawing of vectors to scale and the measurement of angles. More importantly, it does not allow algebraic operations that otherwise would give a simple solution. We can, however, extend algebraic techniques to vectors, *provided vectors are represented on a rectangular coordinate system*. The representation of a vector on a coordinate system uses the concept of unit vector and component.

Now, the stage is set to design a framework, which allows vector addition with algebraic methods. The framework for vector addition draws on two important concepts. The first concept is that a vector can be equivalently expressed in terms of three component vectors:

$$a = a_x i + a_y j + a_z k \quad b = b_x i + b_y j + b_z k$$

The component vector form has important significance. It ensures that component vectors that are to be added are restricted to three known directions only. This paradigm eliminates the possibility of unknown direction. The second concept is that vectors along a direction can be treated algebraically. If two vectors are along the same line, then the resultant is given as :

When $\theta = 0°$, $\cos\theta = \cos 0° = 1$ and,

$$\Rightarrow R = \sqrt{(P^2 + 2PQ + Q^2)} = \sqrt{\{(P+Q)^2\}} = P + Q$$

When $\theta = 180°$, $\cos\theta = \cos 180° = -1$ and,

$$\Rightarrow R = \sqrt{(P^2 - 2PQ + Q^2)} = \sqrt{\{(P-Q)^2\}} = P - Q$$

Thus, we see that the magnitude of the resultant is equal to the algebraic sum of the magnitudes of the two vectors.

Using these two concepts, the addition of vectors is affected as outlined here:

1:  Represent the vectors ($a$ and $b$) to be added in terms of components:

$$a = a_x i + a_y j + a_z k \quad b = b_x i + b_y j + b_z k$$

2:  Group components in a given direction:

$$a + b = a_x i + a_y j + a_z k + b_x i + b_y j + b_z k$$

$$\Rightarrow a + b = (a_x + b_x)i + (a_y + b_y)j + (a_z + b_z)k$$

3:  Find the magnitude and direction of the sum, using the analytical method:

$$\Rightarrow a = \sqrt{\{(a_x + b_x)^2 + (a_y + b_y)^2 + (a_z + b_z)^2\}}$$

## Multiplication with scalar

The multiplication of a vector $A$, with a scalar quantity $a$, results in another vector, $B$. The magnitude of the resulting vector is equal to the product of the magnitude of the vector with the scalar quantity. The direction of the resulting vector, however, is same as that of the original vector (see figures below).

$$B = aA$$

**Multiplication with scalar**

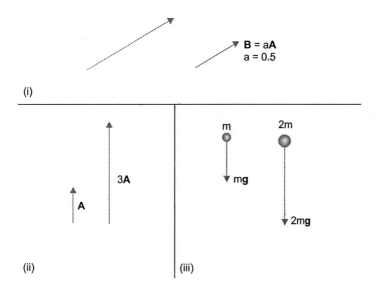

**Figure 3.9**

We have already made use of this type of multiplication intuitively in expressing a vector in component form.

$$A=A_x i+A_y j+A_z k$$

In this vector representation, each component vector is obtained by multiplying the scalar component with the unit vector. As the unit vector has the magnitude of 1 with a specific direction, the resulting component vector retains the magnitude of the scalar component, but acquires the direction of the unit vector.

$$A_x=A_x i$$

# Products of vectors

Some physical quantities are themselves a scalar quantity, but are composed from the product of vector quantities. One such example is "work". On the other hand, there are physical quantities like torque and magnetic force on a moving charge, which are themselves vectors, and are also composed from vector quantities.

Thus, products of vectors are defined in two distinct manners – one resulting in a scalar quantity and the other resulting in a vector quantity. The product that results in scalar value is a scalar product, also known as a dot product. A "dot" (.) is the symbol of the operator for this product. On the other hand, the product that results in vector value is a vector product, also known as a cross product. A "cross" (x) is the symbol of the operator for this product. We shall discuss scalar product only in this module. We shall cover vector product in a separate module.

# Scalar product (dot product)

Scalar product of two vectors $a$ and $b$ is a scalar quantity defined as:

$a.b = ab\cos\theta$

Where "a" and "b" are the magnitudes of two vectors and "$\theta$" is the angle between the direction of two vectors. It is important to note that vectors have two angles $\theta$ and $2\pi - \theta$. We can use either of them, as the cosine of both "$\theta$" and "$2\pi - \theta$" are the same. However, it is suggested to use the smaller of the enclosed angles to be consistent with the cross product in which it is required to use the smaller of the enclosed angles. This approach will maintain consistency with regard to the enclosed angle in two types of vector multiplications.

The notation "$a.b$" is important and should be mentally noted to represent a scalar quantity – even though it involves boldfaced vectors. It should be noted that the quantity on the right hand side of the equation is a scalar.

# Angle between vectors

The angle between vectors is measured with caution. The direction of vectors may sometimes be misleading. The basic consideration is that it is the angle between vectors at the common point of intersection. This intersection point, however, should be the common tail of vectors. If required, we may be required to shift the vector parallel to it or along its line of action to obtain the common point at which tails of vectors meet.

**Angle between vectors**

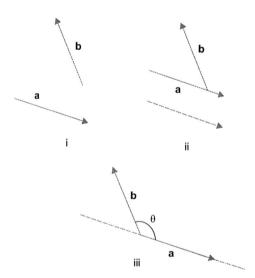

**Figure 3.10: Angle between vectors**

See the steps shown in the figure. First, we need to shift one of two vectors, say $a$, so that it touches the tail of vector $b$. Second, we move vector $a$ along its line of action until the tails of the two vectors meet at the common point. Finally, we measure the angle $\theta$ such that $0 \leq \theta \leq \pi$.

# Meaning of scalar product

We can read the definition of scalar product in either of the following manners :

*a.b=a(bcosθ)*

*a.b=b(acosθ)*

Recall that "bcos θ" is the scalar component of vector *b* along the direction of vector *a*, and "a cos θ" is the scalar component of vector *a* along the direction of vector *b*. Thus, we may consider the scalar product of vectors *a* and *b* as the product of the magnitude of one vector and the scalar component of the other vector along the first vector.

The figure below shows a drawing of scalar components. The scalar component of vectors in figure (i) is obtained by drawing perpendicular from the tip of the vector, *b*, in the direction of vector *a*. Similarly, the scalar component of vectors in figure (ii) is obtained by drawing perpendicular from the tip of the vector *a*, in the direction of vector *b*.

**Scalar product**

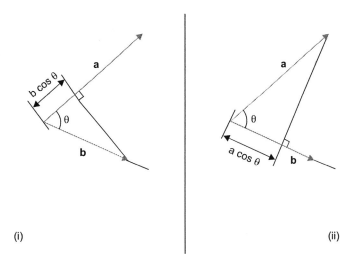

**Figure 3.11**

The two alternate ways of evaluating the dot product of two vectors indicates that the product is commutative i.e. independent of the order of the two vector :

*a.b=b.a*

# Values of scalar product

The value of a dot product is maximized for the maximum value of cosθ. Now, the maximum value of cosine is cos0°=1. For this value, the dot product is simply the product of the magnitudes of the two vectors.

$(a.b)$ max$=ab$

For $\theta=180°$, $\cos180°=-1$ and

$a.b=-ab$

Thus, we see that the dot product can be a negative value. This is a significant result as many scalar quantities in physics are given negative value. The work done, for example, can be negative, when displacement is in the opposite direction to the component of force along that direction.

The scalar product evaluates to zero for $\theta = 90°$ and $270°$ since cosine of these angles is zero. These results have important implications for unit vectors. The dot products of the same unit vector evaluates to 1.

$i.i=j.j=k.k=1$

The dot products of the combination of different unit vectors evaluates to zero.

$i.j=j.k=k.i=0$

## Scalar product in component form

Two vectors in component forms are written as:

$a=a_xi+a_yj+a_zk$

$b=b_xi+b_yj+b_zk$

In evaluating the product, we make use of the fact that multiplication of the same unit vectors is 1, while multiplication of different unit vectors is zero. The dot product evaluates to scalar terms as:

$a.b=(a_xi+a_yj+a_zk).(b_xi+b_yj+b_zk)$

$\Rightarrow a.b=a_xi.b_xi+a_yj.b_yj+a_zk.b_zk$

$\Rightarrow a.b=a_xb_x+a_yb_y+a_zb_z$

## Component as scalar (dot) product

A closer look at the expansion of the dot product of two vectors reveals that the expression is very similar to the expression for a component of a vector. The expression of the dot product is:

$a.b=ab\cos\theta$

On the other hand, the component of a vector in a given direction is:

$a_x=a\cos\theta$

Comparing two equations, we can define the component of a vector in a direction given by unit vector "*n*" as :

$a_x = a.n = a\cos\theta$

This is a very general and useful relation to determine component of a vector in any direction. Only requirement is that we should know the unit vector in the direction in which component is to be determined.

## Attributes of scalar (dot) product

In this section, we summarize the properties of dot product as discussed above. Besides, some additional derived attributes are included for reference.

1. A dot product is commutative:

    This means that the dot product of vectors is not dependent on the sequence of vectors:

    $a.b = b.a$

    We must, however, be careful while writing sequence of dot product. For example, writing a sequence involving three vectors like *a.b.c* is incorrect. For, dot product of any two vectors is a scalar. As dot product is defined for two vectors (not one vector and one scalar), the resulting dot product of a scalar (*a.b*) and that of third vector *c* has no meaning.

2. A dot product is distributive:

    $a.(b+c) = a.b + a.c$

3. The dot product of a vector with itself is equal to the square of the magnitude of the vector:

    $a.a = a \times a\cos\theta = a^2\cos 0° = a^2$

4. The magnitude of the dot product of two vectors can be obtained in either of the following manners:

    $a.b = ab\cos\theta$

    $a.b = ab\cos\theta = a \times (b\cos\theta) = a \times \text{component of } b \text{ along } a$

    $a.b = ab\cos\theta = (a\cos\theta) \times b = b \times \text{component of } a \text{ along } b$

    The dot product of two vectors is equal to the algebraic product of magnitude of one vector and the component of the second vector in the direction of first vector.

5. The cosine of the angle between two vectors can be obtained in terms of a dot product:

    $a.b = ab\cos\theta$

    $\Rightarrow \cos\theta =$

a.b/ab

6.  The condition of two perpendicular vectors in terms of dot product is given by:

$a.b=ab\cos90°=0$

7.  Properties of a dot product with respect to unit vectors along the axes of a rectangular coordinate system are:

$i.i=j.j=k.k=1$

$i.j=j.k=k.i=0$

8.  A dot product in component form is:

$a.b=a_xb_x+a_yb_y+a_zb_z$

9.  The dot product does not yield to cancellation. For example, if $a.b = a.c$, then we cannot conclude that $b = c$. Rearranging, we have:

$a.b-a.c=0$

$a.(b-c)=0$

This means that $a$ and $(b - c)$ are perpendicular to each other. In turn, this implies that $(b - c)$ is not equal to zero (null vector). Hence, $b$ is not equal to $c$, as we would get after cancellation.

We can understand this difference with respect to cancellation more explicitly by working through the problem given here:

## Differentiation and dot product

Differentiation of a vector expression yields a vector. Consider a vector expression given as:

$a=(x^2+2x+3)i$

The derivative of the vector with respect to x is,

$a'=(2x+2)i$

As the derivative is a vector, two vector expressions with a dot product are differentiated in a manner so that the dot product is retained in the final expression of the derivative.

## Projectile Motion

An object that moves near the surface of the earth would normally travel along a parabolic trajectory, unless it is travelling vertically upward or straight in a horizontal line. The object that travels vertically upward will follow the mechanism described under freefall. The object travelling in a horizontal line would follow the last half of a parabolic trajectory until it reaches the ground. In these situations, the acceleration considered should only be the one due to gravity.

In solving projectile motion questions, you need to use the kinematic equations in one dimensions for x and y axes separately. To do this, first resolve the initial velocity to its x and y components. Then, apply the kinematic equations to the x axis and y axis separately.

e.g.

Assume a projectile is launched at an angle of 30 degrees with an initial velocity of 40m/s; the questions asked are:

1.  How far will the projectile travel (this means the range)?

2.  How high will the projectile travel?

3.  What is the total hang-time (this means the time it takes for the projectile to go up and come back down)?

## Step 1:  Resolve the velocity to x and y components;

V ( x) = 40 Cos 30 = 34.64 m/s

V (y)  = 40 Sin 30  = 20 m/s

Now, apply these values to kinematic equations for the x axis and the y axis.

Kinematic equations in the x axis, for a projectile:

Since the horizontal velocity is the same throughout the motion of a projectile (since there is not acceleration and the gravity acts only in the vertical direction), the only equation applicable is:

Range = Horizontal Velocity X time

Time here is the total hang-time.

The time is the only component missing here. Therefore, time has to be calculated by using other kinematic equations applicable in the Y direction.

The kinematic equations in the Y direction:

$V_{(y)\,final} = V_{(y)\,initial} + at$

a = acceleration, but the acceleration here is the gravity acting in the opposite direction. Therefore, the above equation can be re-written as:

$V_{(y)\,final} = V_{(y)\,initial} - gt$

The negative sign indicates that the gravity is acting in the opposite direction.

If you consider the movement of the projectile for a part of the parabolic pathway, then V final is going to be zero. That is when the projectile has reached the highest vertical point. Then the t value is half of the hang-time. If you are going to use the t value to find the range, you need to double the t value.

The other kinematic equations for the y axis, after considering gravity as the acceleration in the opposite direction, are:

$$V^2_{(y\,axis)\,(final)} = V^2_{(initial)} - 2g\,\Delta y$$

Y is the distance travelled in the vertical direction.

$$\Delta y = (v_{(y)\ initial}\qquad {}^{2}$$

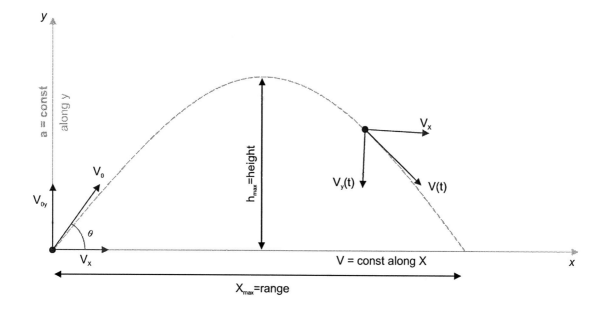

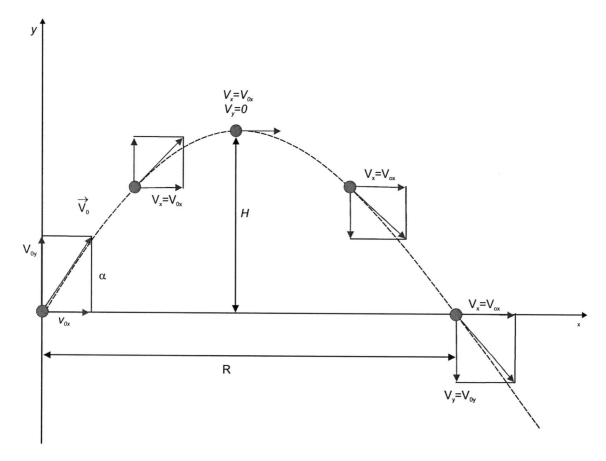

**Figure 3.12 explaining the changes in velocity in horizontal and vertical directions during a projectile motion. The horizontal motion remains a constant.**

For questions and answers on vectors, please visit the website:

http://fountain.cnx.rice.edu:8280/content/m13601/latest/

For questions and answers on projectile motions, please visit the website:

http://cnx.org/content/m13866/latest/

# PowerPoint Link:

Please refer to the end of the module lecture links.

# Discussion Question

Answer the following questions with references. Please remember to follow the standard APA referencing style.

For APA standards of references, please visit: http://owl.english.purdue.edu/owl/resource/560/01/

Also, respond in detail to one other post by fellow students.

3.1 "All roads lead to Rome" - analyze this saying in terms of vectors and displacement, and give examples.

 3.2 Describe some important changes you would suggest to a "trebuchet" manufacturer. Explain your improvement in terms of motion in two dimensions.

# Laboratory Activity and the link:

Go to the link below and run the simulation:

http://phet.colorado.edu/simulations/sims.php?sim=Projectile_Motion

Go to: http://phet.colorado.edu/teacher_ideas/view-contribution.php?contribution_id=809

Open Projection Motion.doc

Answer all the questions.

<u>Now take the chapter 3 test (not included with this book)</u>

# Module 1 Lecture Links:

http://ocw.mit.edu/courses/physics/8-01-physics-i-classical-mechanics-fall-1999/video-lectures/lecture-1/

http://ocw.mit.edu/courses/physics/8-01-physics-i-classical-mechanics-fall-1999/video-lectures/lecture-1/

http://ocw.mit.edu/courses/physics/8-01-physics-i-classical-mechanics-fall-1999/video-lectures/lecture-3/

http://ocw.mit.edu/courses/physics/8-01-physics-i-classical-mechanics-fall-1999/video-lectures/lecture-4/

# Module 1 - Student's self-assessment

Please answer the following questions, which give you an indication of your standard of learning for this module:

- Do you know how to perform dimensional analysis?
- Do you understand significant figures?
- Do you know the difference between distance and displacement?
- Do you know how to define velocity and speed?
- Do you know how to define acceleration?
- Can you plot various graphs including distance, displacement, velocity, acceleration and time?
- Can you carry out calculations using basic equations for objects moving in one dimension?
- Can you describe vectors and scalars?
- Do you know the terms instantaneous velocity and instantaneous acceleration?
- Can you apply your knowledge to solve problems on free falling objects?
- Do you know how to perform two dimensional analysis?
- Do you know the components of a projectile motion?
- Can you find out the resultant of two or more vectors?
- Can you carry out calculations using basic equations for objects moving in two dimensions?
- What sections of the textbooks would you read for more information on the above?
- Did you enjoy your lesson?
- What aspect of the lesson was most interesting to you?
- Can you now confidently do your activities?
- What aspect of the lesson was most interesting to you?
- Can you now confidently do the lesson's activities?

# Module 2

## CHAPTER-4

# The Laws of Motion

## Objectives

At the end of this lesson, you should be able to:

1.  Explain Newton's laws;

2.  Apply Newton's laws to solve problems.

## Lecture Notes

### Mass and the weight

Mass is a scalar. It has only magnitude. The total mass is given by the mass of the total atoms in a body. The SI unit for mass is kg. When a body is at rest, relative to a given frame of reference, it is known as the rest mass. If a body is moving near to the speed of light, then according to Einstein's theory of relativity, the mass will change. Nevertheless, under normal conditions, where the speed of the object does not travel near the speed of light, the mass is constant.

Mass could also be described as a measure of the inertia of a body. The more mass a body has, the more difficult it is to change its state of motion if it is moving, or to move it if it is at rest.

The gravitational force exerted on mass is known as the weight. This in mathematical terms is mass times the gravitational acceleration.

The weight force = $F_w$ = mg. This is always directed perpendicularly to the Earth's surface. The SI unit for weight is N.

# Normal force

This is the upward force exerted by the surface on an object lying on the surface. This force is always perpendicular to the surface.

If the surface is horizontal, then the object lying on it is exerting its weight force on the surface. The surface in turn is exerting an equal and opposite normal force on the object.

# Friction Force

There are two types of frictional forces. They are the static frictional force and the kinetic frictional force. The static frictional force is the resistance to movement exerted by surfaces on the unmoving objects lying on them. When an object is sliding along the surface, then the force exerted by the surface against the sliding in the opposite direction to the movement of the object is known as the kinetic frictional force. The coefficient of friction (either static or kinetic) determines the magnitude of the force. The static frictional force is at its maximum just before the object begins to slide. When the object is sliding, the kinetic frictional force is applied. This is always less than the maximum static frictional force.

(Fk) Frictional force (kinetic) = (Fn)Normal force  X Coefficient of kinetic frictional force ($\mu$)

(Fs) Frictional force (static) = Normal force  X Coefficient of static frictional force

In a horizontal plane, the normal force is equal to the weight of the object.

Therefore, the normal force = mg

On an inclined plane, making an angle $\theta$ between the plane and the horizontal surface, the normal force is equal to the portion of the weight exerted on the surface of the inclined plane. That portion of the weight is given as:

Mg Cos $\theta$

Therefore,

F (k) = mg Cos $\theta$ X $\mu$

When the object is sliding down the inclined plane, the force acting in the direction of the movement is a portion of the weight. This is equal to mg Sin $\theta$.

When this force is matched by the kinetic frictional force acting in the opposite direction, it is equal to mg Sin $\theta$. The object is then moving at a constant velocity (see Newton's second law).

When these forces are not balanced the object will accelerate down the plane.

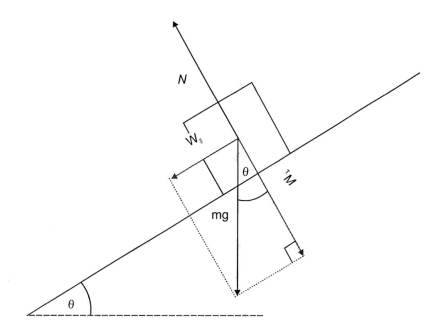

**Figure 4.1: The components of weight acting in x and y directions.**

Wll is the mg Sin θ and W⊥ is mg Cos θ

# Newton's Laws

## Newton's first law

### Definition:

A body continues in its state of rest or of uniform speed in a straight line as long as no net force acts on it.

Any object in general resists change. This resistance to change is known as inertia. If an object is moving, it will resist the forces trying to stop its movement; if an object is not moving (at rest), then it will resist the forces trying to make it move. This resistance is inherent to its mass. The more mass an object possesses, the more inertia it displays.

Newton's laws are valid in an inertial frame of reference. This means, if the frame of reference is moving or accelerating, then Newton's laws would not be valid. Any object that moves at constant velocity relative to an inertial frame of reference is also an inertial frame of reference.

Net force exists when the opposing forces are not balanced. These forces can be in any dimension (e.g. x, y, and z).

(Read the difference between mass and weight)

e.g.

A book lying on a table:  The weight of the book, which is vertically downward (the weight force,) and the upward force by the table-top on the book, which is vertically upward (the normal force), are equal. Therefore, there is no net force. As a result, the book remains at rest.

A car moving at a constant velocity: The forces acting on the car are balanced. The forward applied force that moves the car forward and the backward forces, the friction, and the air resistance, are balanced. Therefore, there is no net force. These are horizontal forces. Similarly, the vertical forces, the weight of the car and the normal force acted on the car by the road, are balanced i.e. the car is not moving up and down in the vertical direction. The fact that the moving object is at constant velocity indicates that there is no net force.

The mass here is inertial mass or gravitational mass, not relativistic mass. This is because the mass is considered under a frame of reference where there is no change (i.e. inertial frame of reference).

When we are seated on a chair, we are at rest. If we analyze this further, we will realize that we are truly not at rest. The earth is rotating around the sun. Therefore, everything on earth too rotates with the earth. Therefore, our statement of "at rest" is relative to a frame of reference. The application of Newton's first law is based on a frame of reference. Since Galileo discovered this first, the reference frame is called the Galilean Reference. This is an inertial frame of reference.

Within a frame of reference, if an object is at rest or in motion, to change that state, we need to create a net force. This means to create an imbalance of forces in the direction of interest. Therefore, the force is what changes the state of rest or uniform motion.

The motion of a body is dependent on the speed (or velocity) and its mass. Therefore, mass, as described earlier, resists the change.

In order to understand if a body is at rest or in motion, first check if there is a net force acting on it. If there is no net force, then the object is either at rest or moving at a constant velocity. If there is a net force, then the object at rest would move, and the object that is in motion will accelerate or decelerate (acceleration in the opposite direction).

# Demonstration:

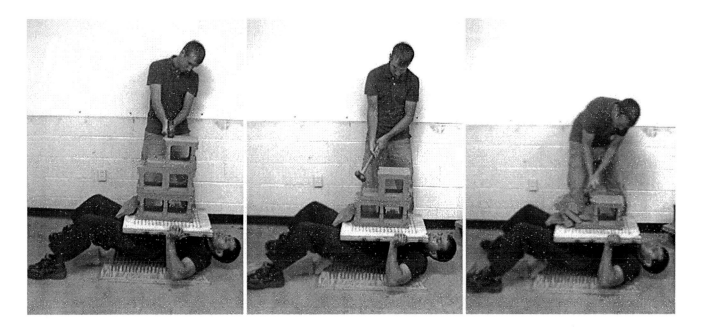

**Figure 4.2 : A Martial Arts demonstration: A man is sandwiched between two beds of nails, and three concrete blocks are placed on top as shown in the picture. Another man is breaking the blocks with a hammer.**

Describe the reason why the man sandwiched between the two beds of nails was unharmed in the demonstration.

The substance of the first law of motion is expressed in many ways. Here, we sum them all for ready reference (for the condition that the net force on a body is zero):

- The body may either be at rest or may move with constant velocity.

- The body is not associated with any acceleration.

- If the body is moving, then the body moves along a straight line with a constant speed without any change of direction.

- If the body is moving, then the motion of the body is a uniform linear motion.

- If a body is moving with uniform linear motion, then we can be sure that the net force on the body is zero.

An inertial frame of reference, where Newton's first law is valid, moves with constant velocity without acceleration. A frame of reference moving with constant relative velocity with respect to an inertial frame of reference is also an inertial frame of reference. The context of inertial frame of reference is important for Newton's first law. Otherwise, we may encounter situations where a body may be found to be accelerated, even when the net force acting on the body is zero. Consider the book lying on the floor of an accelerated lift that is moving upward.

### Books lying on the floor of an accelerated lift

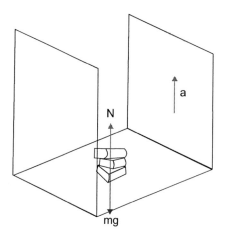

**Figure 4.3: The book is acted upon by a pair of balanced forces.**

To an observer in the lift, the books are under a pair of balanced forces: the weight of the books acting downward and an equal normal force acting upward. The net force on the book is zero. An observer on the ground, however, finds that the book is accelerated up. To support this observation in Earth's inertial frame, the observations in the lift have to be incorrect.

This apparent paradox is resolved by restraining ourselves to apply Newton's first law in the inertial frame of reference only. The observer on the ground determines that the book is moving with upward acceleration. He concludes that the normal force is actually greater than the weight of the books such that

**Books lying on the floor of an accelerated lift**

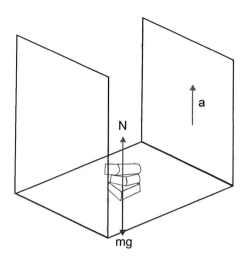

**Figure 4.4: The book is acted upon by a net force.**

N - mg = ma

Alternatively, we may use the technique of pseudo force and convert the accelerated frame into an inertial frame and then apply Newton's first law. We shall discuss this technique subsequently.

Further, we can always convert an accelerated non-inertial frame of reference to an equivalent inertial frame of reference, using the concept of "pseudo force". This topic will be dealt with in detail separately.

We now sum up the discussion thus far as:

- Inertial frame of reference is one in which Newton's first law of motion is valid.

- Inertial frame of reference is one which moves at uniform velocity.

- Any reference system, which is moving with uniform velocity with respect to an inertial frame of reference, is also an inertial frame of reference.

- Earth's frame of reference approximates to inertial frame for motion, which is limited in dimension.

- We can convert an accelerated non-inertial frame of reference to an equivalent inertial frame of reference, using the concept of "pseudo force".

# Exercise

In which case(s) do you see the net force as non-zero ?

(a)   An air bubble moving up inside a soda bottle at a speed 0.1 m/s

(b)   A cork floating on water

(c)   A car moving at 60 km/hr on a rough horizontal road

(d)  None of above

## Solution

The objects in cases (a) and (c) are moving with constant velocity. Thus, there is no net force in these cases. On the other hand, floatation results when net force is zero.

Hence, option (d) is correct.

# Newton's second law

The rate of change of the momentum of a moving body is directly proportional to the net (resultant) external force acting upon it.

The momentum (p) is defined (discussed under another chapter) as velocity times mass.

$$p = mv$$

In mathematical terms,

$$F = dp / dt$$

$$F = d(mv)/dt$$

For invariant mass (the mass of the body under consideration does not change during application of force), we can take out mass "m" from the differential sign:

$$F = M (dv/dt)$$

$$F = mv/t$$

Dv/dt is the acceleration.

Hence,

$$F = ma$$

Here the F is the net force. The net force left after balancing the forces acting on a particular direction of movement (i.e. either horizontal or vertical directions).

A smaller mass yields a greater acceleration and a greater mass yields a smaller acceleration. Thus, "mass" of a body is the measure of the inertia of the body in translational motion. Here, we have specified that mass is the measure of inertia in translational motion, because the inertia to rotational motion is measured by a corresponding rotational term called "moment of inertia".

The second law of motion determines the effect of net force on a body. The first law only defines the natural state of the motion of a body, when net force on the body is zero. It does not provide us with any tool to quantitatively relate force and acceleration (rate of change in velocity).

The second law of motion is the centerpiece of classical dynamics as it states the exact relation between force (cause) and acceleration (effect). This law has an explicit mathematical form and, therefore, has the advantage of quantitative measurement. In fact, the only available quantitative definition of force is given in terms of the second law: "Force is equal to acceleration produced in unit mass."

It must be clearly understood that the three laws of motion could well have been replaced by this single law of motion. However, the three laws are presented as they are, because the first and third laws convey the fundamental nature of "motion" and "force", which are needed to complete our understanding about them.

The second law of motion is stated in terms of linear momentum. Therefore, it would be appropriate that we first familiarize ourselves with this term.

Linear momentum of a particle is defined as a vector quantity, having both magnitude and direction. It is the product of the mass (a scalar quantity) and the velocity (a vector quantity) of a particle at a given instant.

$p = \mathrm{m}v$

The dimensional formula of linear momentum is $[MLT^{-1}]$ and its SI unit of measurement is "$kg{-}m/s$ ".

A few important aspects of linear momentum need our attention:

First, linear momentum is a product of a positive scalar (mass) and a vector (velocity). This means that the linear momentum has the same direction as that of velocity.

The motion of a body could be represented completely by velocity. However, the velocity alone does not convey anything about the inherent relation that "change in velocity" has with force. The product of mass and velocity in linear momentum provides this missing information.

In order to fully appreciate the connection between motion and force, we may consider two balls of different masses, moving at the same velocity, which collide with a wall. It is our everyday common sense that tells us that the ball with greater mass exerts more force on the wall. We may therefore conclude that linear momentum, i.e. the product of mass and velocity, represents the "quantum of motion", which can be connected to force.

It is this physical interpretation of linear momentum that explains why Newton's second of motion is stated in terms of linear momentum, as this quantity (not the velocity alone) connects motion with force.

A few important aspects of Newton's second law of motion are discussed in the following section:

## Deduction of first law of motion

Newton's first law of motion is a subset of the second law in the sense that the first law is just a specific description of motion, when net external force on the body is zero.

$F=ma= 0$

$a = \mathrm{m}\ (dv/dt) = 0$

This means that if there is no net external force on the body, then acceleration of the body is zero. Equivalently, we can state that if there is no net external force on the body, then the velocity of the body cannot change. It is also easy to infer that if there is no net external force on the body, it will maintain its state of motion. These are exactly the statements in which the first law of motion is stated.

# Forces on the body

A body under investigation may be acted upon by a number of forces. We must use the vector sum of all external forces, while applying the second law of motion. A more general form of the second law of motion, valid for a system of force is:

$$\sum F = ma$$

The vector addition of forces as required in the left hand side of the equation excludes the forces that the body applies on other bodies. We shall know from the third law of motion that force always exists in the pair of "action" and "reaction". Hence, if there are "n" numbers of forces acting on the body, then there are "n" numbers of forces that the body exerts on other bodies. We must carefully exclude all such forces that the body applies on other bodies as a reaction.

Consider two blocks "A" and "B" lying on a horizontal surface as shown in the figure. There are many forces in action here:

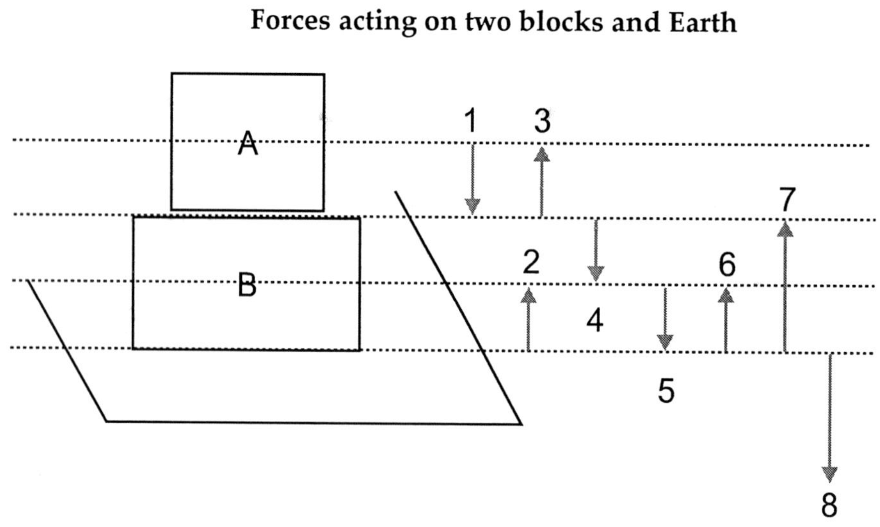

**Forces acting on two blocks and Earth**

**Figure 4.5: There are many forces acting on different bodies.**

1.  Earth pulls down "A" (force of gravitation)

2.  "A" pulls up Earth (force of gravitation)

3.  "B" pushes up "A" (normal force)

4.  "A" pushes down "B" (weight of "A" : normal force)

5.  Earth pulls down "B" (force of gravitation)

6.  "B" pulls up Earth (force of gravitation)

7.  Earth pushes up "B" (normal force) and

8.  "B" pushes down the Earth (weight of "B" and "A" : normal force)

Let us now consider that we want to apply the law of motion to block "B". There are a total of six forces related to "B": 3, 4, 5, 6, 7 and 8. Of these, three forces (4, 5 and 7) act on "B", while the remaining equal numbers of forces are applied by "B" on other bodies.

**External forces acting on block B**

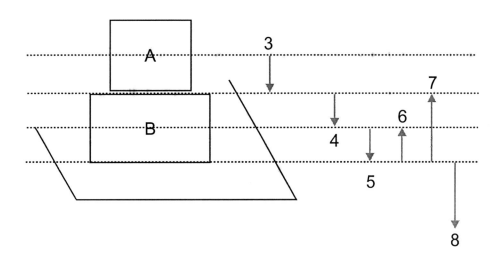

**Figure 4.6: There are three external forces on B as shown with red arrow.**

From the point of view of the law of motion, there are, thus, only three forces as shown with red arrows in the figure, which are external forces. The surrounding applies these forces on the block "B".

As pointed out before in the course, we must not include internal forces like intermolecular forces in the consideration. In order to understand the role and implication of internal forces, we consider the interaction of two blocks together with Earth's surface. In this case, the forces 3 (B pushes up A), and 4 (A pushes down B), are a pair of normal forces acting at the interface between blocks A and B. They are internal to the combined system of two bodies and as such are not taken into account when using the law of motion. Here, external forces are 1 (Earth pulls down A), 5(Earth pulls down B), and 7(Earth pushes up B).

**External and internal forces acting on blocks A and B**

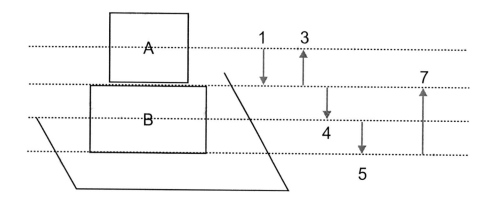

**Figure 4.7: External and internal forces acting on blocks A and B**

In nutshell, we must exclude (i) forces applied by the body and (ii) internal forces. Clearly, we must only consider external forces applied on the body while using the equation of motion as given by Newton's second law.

# Equation of motion in component directions

As the force and acceleration are vector quantities, we can represent them with three components in mutually perpendicular directions. The consideration of dimension is decided by the force system as applied to the body.

If forces are collinear i.e. acting along a particular direction, then we use the equation of motion in one direction. In such a situation, it is possible to represent vector quantities with equivalent signed scalar quantities, in which the sign indicates the direction. This is the simplest case.

If forces are coplanar i.e. acting in a plane, then we use component equations of motion in two directions. In such a situation, we use component equations of motion in two directions.

$\sum F_x = ma_x$

$\sum F_y = ma_y$

Since each coordinate direction is bi-directional, we treat component vectors by equivalent scalar representation whose sign indicates direction. The important aspect of component equation of motion is that acceleration in a particular component direction is caused by the net of force components in that direction, and is independent of other net of component force in the perpendicular direction.

In cases where the forces are distributed in three-dimensional volume, then we must consider the component equation of motion in the third perpendicular direction also:

$\sum F_z = ma_z$

# Application of force

We implicitly consider that forces are applied at a single point object. The forces that act on a point object are concurrent by virtue of the fact that a point is a dimensionless entity. This may appear to be confusing, as we have actually used the word "body" – not "point", in the definition of the second law. Here, we need to appreciate the intended meaning clearly.

Actually, the second law is defined in the context of translational motion, in which a three dimensional real body behaves like a point. We shall subsequently learn that application of a force system (forces) on a body in translation is equivalent to a point, where all mass of the body can be considered to be concentrated. In that case, the acceleration of the body is associated with that point, which is termed as "center of mass (C)". Newton's second law is suitably modified as:

$$\sum F = ma_c$$

where $a_c$ is the acceleration of the center of mass (we shall elaborate about the concept of center of mass in a separate module). In general, application of a force system on a real body can involve both translational and rotational motion. In such a situation, the concurrency of the system of forces with respect to points of application on the body assumes significance. If the forces are concurrent (meeting at a common point), then the force system can be equivalently represented by a single force, applied at the common point. Further, if the common point coincides with "center of mass (C)", then the body undergoes pure translation. Otherwise, there is a turning effect (angular/rotational effect) that is also involved.

(a) Common point coincides with center of mass

(b) Common point does not coincide with center of mass

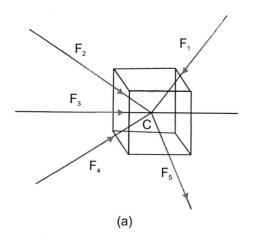
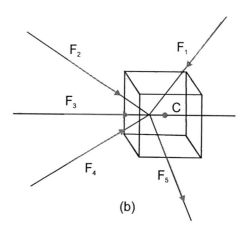

(a)                                    (b)

**Figure 4.8: Concurrent force system**

Non-concurrent force system

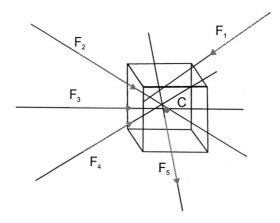

**Figure 4.9: Forces as extended do not meet a common point**

$$\tau = |r \times F| = r_\perp F$$

where $r_\perp$ is the perpendicular distance from the point of rotation.

Most importantly, same force or force system is responsible for both translational effect (force acting as "force" as defined by the second law) and angular/rotational effect (force manifesting as "torque" as defined by the angular form of Newton's second law in the module titled <u>Second law of motion in angular form</u> ). We leave the details of these aspects of the application of force, as we will study it separately. However, the point is made. Linear acceleration is not the only "effect" of the application of force (cause).

Also, force causes "effect" not necessarily as cause of acceleration – but can manifest in many ways: as torque to cause rotation; as pressure to change volume, as stress to deform a body, etc. We should, therefore, always keep in mind that the study of the translational effect of force is specific and not inclusive of other possible effects of force(s).

In the following listing, we intend to clarify the context of the study of the motional effect of force:

1.    The body is negligibly small to approximate a point. We apply Newton's second law for translation as defined without any consideration of turning effect.

2:    The body is a real three dimensional entity. The force system is concurrent at a common point. This common point coincides with the center of mass. We apply Newton's second law for translation as defined without any consideration of turning effect. Here, we implicitly refer the concurrent point as the center of mass.

3:    The body is a real three dimensional entity. The force system is concurrent at a common point. However, this common point does not coincide with the center of mass. The context of study in this case is also the same as that for the case in which the force system is not concurrent. We apply Newton's second law for translation as defined for the center of mass and Newton's second law for angular motion (we shall define this law at a later stage) for angular or rotational effect.

The discussion so far assumes that the body under consideration is free to translate and rotate. There are, however, real time situations in which rotational effect due to external force is counter-balanced by restoring torque. For example, consider the case of a sliding block on an incline. Application of an external force on the body along a line, not passing through the center of mass, may not cause the body to overturn (rotate as it moves). The moment of external force i.e. applied torque, may not be sufficient enough to overcome restoring torque due to gravity. As such, if it is stated that the body is only translating under the given force system, then we assume that the body is a point mass, and we apply Newton's second law straight away as if the body were a point mass.

Unless otherwise stated or specified, we shall assume that the body is a point mass and forces are concurrent. We shall, therefore, apply Newton's second law, considering forces to be concurrent, even if they are not. Similarly, we will consider that the body is a point mass, even if it is not. For example, we may consider a block, which is sliding on an incline. Here, "friction force" is along the interface, whereas the "normal force" and "weight" of the block act through the center of mass (C). Obviously, these forces are not concurrent. However, we apply Newton's second law for translation, as if forces were concurrent.

# Exercise

## Exercise 1

A ball of mass 0.1 kg is thrown vertically. In which of the following case(s) is the net force on the ball zero? (ignore air resistance)

a. Just after the ball leaves the hand

b. During upward motion

c. At the highest point

d. None of above cases

*Solution*

The only force acting on the ball is gravity (gravitational force due to Earth). It remains constant as acceleration due to gravity is constant in the vicinity of Earth. Thus, net force on the ball during its flight remains constant, which is equal to the weight of the ball i.e.

$F=mg=0.1 \times 10=1N$

Hence, option (d) is correct.

# Exercise 2

A pebble of 0.1 kg is subjected to different sets of forces in different conditions on a train, which can move on a horizontal linear direction. Determine the case when magnitude of net force on the pebble is greatest. Consider g = 10 $m/s^2$ .

a. The pebble is stationary on the floor of the train, which is accelerating at 10 $m/s^2$ .

b. The pebble is dropped from the window of the train, which is moving with uniform velocity of 10 m/s.

c. The pebble is dropped from the window of the train, which is accelerating at 10 $m/s^2$ .

d. The magnitude of force is equal in all the above cases.

*Solution*

In case (a), the pebble is moving with horizontal acceleration of 10 m/s² as seen from the ground reference (inertial frame of reference). The net external force in the horizontal direction is :

$F_{net}= ma =0.01 \times 10 =1N$ (acting horizontally)

There is no vertical acceleration. As such, there is no net external force in the vertical direction. The magnitude of net force on the pebble is, therefore, 1 N in horizontal direction.

In case (b), the pebble is moving with a uniform velocity of 10 m/s in the horizontal direction as seen from the ground reference (inertial frame of reference). There is no net force in the horizontal direction. When dropped, the only force acting on it is due to gravity. The magnitude of net force is, thus, equal to its weight :

$F_{net}= mg = 0.01 \times 10 = 1N$ (acting downward)

In case (c), the pebble is moving with an acceleration of 10 $m/s^2$ as seen from the ground reference (inertial frame of reference). When the pebble is dropped, the pebble is disconnected from the accelerating train. As force has no past or future, there is no net horizontal force on the pebble. The only force acting on the pebble is its weight. The magnitude of force on the pebble is:

$F_{net}= mg = 0.01 \times 10 = 1N$ (acting downward)

Hence, option (d) correct.

# Exercise 3

A rocket weighing 10000 kg is blasted upwards with an initial acceleration 10 $m/s^2$ . The thrust of the blast is:

a. $1 \times 10^2$

b. $2 \times 10^5$

c. $3 \times 10^4$

d. $4 \times 10^3$

## Solution

The net force on the rocket is the difference of the thrust and weight of the rocket (thrust being greater). Let thrust be F, then applying Newton's second law of motion:

$$F_{net} = F - mg = ma$$
$$F = m(g+a) = 10000 \times (10+10) = 200000 = 2 \times 10^5 N$$

Hence, option (b) is correct.

# Exercise 4

The motion of an object of mass 1 kg along the x-axis is given by the equation,

$$x = 0.1t + 5t^2$$

where "x" is in meters and "t" is in seconds. The force on the object is:

(a)$0.1N$  (b)$0.5N$  (c)$2.5N$  (d)$10N$

## Solution

The given equation of displacement is a quadratic equation in time. This means that the object is moving with a constant acceleration. Comparing the given equation with the standard equation of motion along a straight line:

$$x = v_i t + \tfrac{1}{2} at^2$$

We have the acceleration of the object as:

$$a = 5 \times 2 = 10 m/t^2$$

Applying Newton's second law of motion:

$$F = ma = 1 \times 10 = 10N$$

Hence, option (d) is correct.

# Exercise 5

A bullet of mass 0.01 kg enters a wooden plank with a velocity 100 m/s. The bullet is stopped at a distance of 50 cm. The average resistance by the plank to the bullet is,

a    $10N$

b.    100N

c.    1000N

d.    10000N

## Solution

The plank applies resistance as force. This force decelerates the bullet and is in the opposite direction to the motion of bullet. Since we seek to know average acceleration, we shall consider this force as a constant force, assuming that the wooden plank has uniform constitution. Let the corresponding deceleration be "a". Then, according to the equation of motion,

$$Vi^2 = Vf^2 + 2ax$$

$$a = -(Vf^2 - Vi^2)/2x$$

Putting values,

$$a = -(10^4 - 0)/(2 \times 0.5) = -10^4 \, m/s^2$$

# Newton's third law

One body interacts with another body exerting force on each other, which is equal in magnitude, but opposite in direction.

The action and reaction pair acts along the same line. Their points of application are different as they act on different bodies. This is a distinguishing aspect of the third law with respect to the first two laws, which consider application of force on a single entity.

The law underlines the basic manner in which force comes into existence. Force results from the interaction of two bodies, always appearing in a pair. In other words, the existence of a single force is impossible. In the figure below, we consider a block at rest on a table. The block presses the table down with a force equal to its weight (mg). The horizontal table surface, in turn, pushes the block up with an equal normal force (N), acting upwards.

## Newton's third law of motion

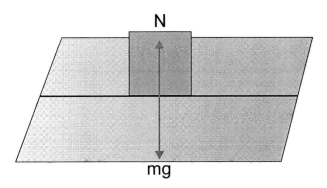

**Figure 4.10: Two bodies exert equal but opposite force on each other.**

$N$=mg

In this case, the net force on the block and table is zero. The force applied by the table on the block is equal and opposite to the force due to gravity acting on it. As such, there is no change in the state of the block. Similarly, net force on the table is zero as the ground applies upward reaction force on the table to counterbalance the force applied by the block. We should, however, be very clear that the action and reaction force arising from the contact are capable of changing the state of motion of individual bodies, provided they are free to move. Consider the collision of two billiard balls. The action and reaction forces during collision changes the course of motion (acceleration of each ball).

The "action" and "reaction" forces are external forces on individual bodies. Depending on the state of a body (i.e. the state of other forces on the body), the individual "action" or "reaction" will cause acceleration in the particular body. For this reason, the book and table do not move on contact, but balls after collision actually move with certain acceleration.

The scope of this force is not limited to interactions involving physical contact. This law appears to apply only when two bodies come in contact. In reality, the characterization of force by the third law is applicable to all force types. This requirement of pair existence is equally applicable to forces like electrostatic or gravitational force, which act at a distance without coming into contact.

$F = q1\ q2\ /\ 4\pi\varepsilon_0 r^2$

The charge $q_1$ applies a force $F_{21}$ on the charge $q_2$ and charge $q_2$ applies a force $F_{12}$ on $q_1$ . The two forces are equal in magnitude, but opposite in direction such that :

**Electrostatic force**

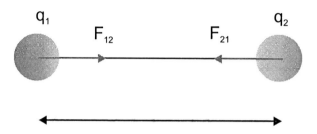

**Figure 4.11: Force appears in pair.**

$$F_{12}=-F_{21}$$

$$F_{12}+F_{21}=0$$

Here, we read the subscripted symbol like this : $F_{12}$ means that it is a force, which is applied on body 1 by body 2.

It should be emphasized that although the vector sum of the two forces is zero, this condition does not indicate a state of equilibrium. This is so because two forces, often called action and reaction pair, are acting on different bodies. Equilibrium of a body, on the other hand, involves consideration of external forces on the particular body.

We have pointed out that "action" and "reaction" forces are external forces on the individual bodies. However, if we consider two bodies forming a "system of two bodies", then action and reaction pairs are internal to the system of two bodies. The forces on the system of bodies are:

$$\sum F=\sum F_{int}+\sum F_{ext}$$

If no external forces act on the system of bodies, then:

$$\sum F_{ext}=0$$

From the second law of motion, we know that only external force causes acceleration to the body system under consideration. As such, acceleration of the "system of two bodies" due to net internal forces should be zero. Hence,

$$\sum F_{int}=0$$

This is possible when internal forces are pair forces of equal magnitude, which are directed in opposite directions.

The internal forces are incapable to produce acceleration of the system of bodies. The term "system of bodies" is important (we shall discuss the concept of system of bodies and their motion in a separate module). The acceleration of the system of bodies is identified with a point known as the center of mass. When we say that no acceleration is caused by the pair of third law forces, we mean that the "center of mass" has no acceleration. Even though the individual body of the system is accelerated, the "center of mass" is not accelerated and hence, we say that the "system of bodies" is not accelerated.

Referring to a two-charged-body system that we to referred earlier, we can consider forces $F_{12}$ and $F_{21}$ being the internal forces with respect to the system of two charged bodies. As such, applying Newton's second law,

# Gravitational force

The force of gravitation is a long distance force, arising due to the very presence of matter. Netwon's gravitation law provides the empirical expression of gravitational force between two point-like masses $m_1$ and $m_2$ separated by a distance "r" as,

$$Fg = G\, m_1\, m_2\, /\, r^2$$

where "G" is the universal constant $G=6.7 \times 10^{-11} N{-}m^2/kg^2$ .

Gravitational force is a pair of pulling forces on the two bodies directed towards each other. It is always a force of attraction. Gravitation is said to follow the inverse square law as the force is inversely proportional to the square of the distance between the bodies.

Since the force of gravitation follows inverse square law, the force can be depicted as a conservative force field, in which work done in moving a mass from one point to another is independent of the path followed. The gravitation force is the weakest of all fundamental forces but can assume great magnitude as there are truly massive bodies present in the universe.

In the case of Earth (mass "M") and a body (mass "m"), the expression for the gravitational force is:

$$Fg = GMm/r^2$$

$$Fg = mg$$

where "g" is the acceleration due to gravity.

$$g = GM\, /\, r^2$$

The most important aspect of acceleration due to gravity here is that it is independent of the mass of the body "m", which is being subjected to acceleration. Its value is taken as 9.81 $m/s^2$ .

Gravitational force has a typical relation with the mass of the body on which its effect is studied. We know that mass ("m") is part of the Newton's second law that relates force with acceleration. Incidentally, the same mass of the body ("m") is also a part of the equation of gravitation that determines force. Because of this special condition, bodies of different masses are subjected to the same acceleration. Such is not the case with other forces and the resulting acceleration is not independent of mass.

We see here that gravitational force on the body is proportional to the mass of the body itself. As such, the acceleration, which is equal to the force divided by mass, remains same.

Knowing that acceleration due to gravitational force in the Earth's vicinity is a constant, we can calculate gravitational pull on a body of mass "m", using the relation of the second law of motion:

$$F = mg$$

Further, gravitational force is typically a force which operates at a distance. The force is said to be communicated to an object at a distance through a field known as a gravitational field. For this reason, gravitational force is classified as "force at a distance".

Forces differ in the manner they operate on a body. Some need physical contact, whereas others act from a distance. Besides, there are forces that apply along the direction of attachments to the body. We can classify forces on the basis of how they operate on a body. According to this consideration, the broad classifications are as given here:

- Field force

- Contact force

- Force applied through an attachment

Forces like electromagnetic and gravitational forces act through a field and apply on a body from a distance without coming into contact. These forces are said to transmit through the field at the speed of light. These forces are known as field forces.

Forces such as normal and friction forces arise when two bodies come into contact. These forces are known as contact forces. They operate at the interface between two bodies and act along a line, which forms a specific angle to the tangent drawn at the interface.

Besides, some of the forces are applied through flexible entities like strings and springs. For example, we raise a bucket from a well by applying force on the bucket through the flexible medium of a rope. The direction of the taut rope (string) determines the direction of force applied. Application of force through a string provides tremendous flexibility as we can change string force direction by virtue of some mechanical arrangement like a pulley.

The spring force is a mechanism through which variable force ($F = kx$) can be applied to a body. Like string force, spring force also provides for changing direction of force by changing the orientation of the spring.

## Contact force

When an object is placed on another object, the two surfaces of the bodies in contact apply normal force on each other. Similarly, when we push a body over a surface, the force of friction, arising from electromagnetic attraction among molecules at the contact surface, opposes the relative motion of two bodies.

In dynamics, we come across these two particular contact forces as described here.

## Normal force

When two bodies come in contact, they apply equal and opposite forces on each other in accordance with Newton's third law of motion. As the name suggest, this "normal force" is normal to the common tangent drawn between two surfaces.

We consider here a block of mass "m" placed on the table. The weight of the block (mg) acts downward. This force tries to deform the surface of the table. The material of the table responds by applying force to counteract the force that tries to deform it.

**Normal force**

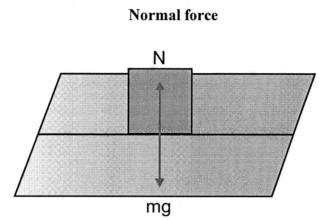

**Figure 4.12: The surface applies a normal force on the body**

The rigid bodies counteract any deformation in equal measure. A rigid table, therefore, applies a force, which is equal in magnitude and opposite in direction. As there is no relative motion at the interface, this contact force has no component along the interface and as such it is normal to the interface.

N=mg

# Friction force

The force of friction comes into play whenever two bodies in contact either tend to move or actually move. The surfaces in contact are not a plane surface as they appear to be. Their microscopic view reveals that they are actually uneven with small hills and valleys. The bodies are not in contact at all points, but limited to elevated points. The atoms/molecules constituting the surfaces attract each other with electrostatic force at the contact points and oppose any lateral displacement between the surfaces.

**Friction force**

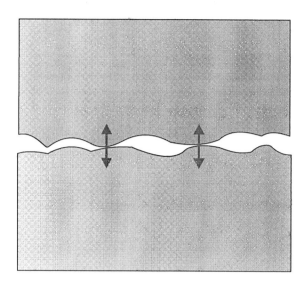

**Figure 4.13: The surfaces are uneven**

For this reason, we need to apply a certain external force to initiate motion. As we increase force to push an object on a horizontal surface, the force of friction also grows to counteract the push that tries to initiate motion. However, beyond a point, when applied force exceeds maximum friction force, the body starts moving.

**Friction force**

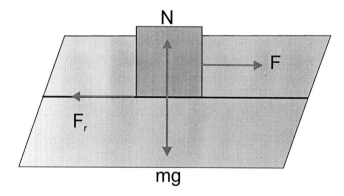

**Figure 4.14: The friction force acts tangential to the interface between two surfaces.**

Incidentally, the maximum force of friction, also known as limiting friction, is related to the normal force at the surface:

$F = \mu N$

where "μ" is the coefficient of friction between two particular surfaces in question. Its value is dependent on the nature of surfaces in contact and the state of motion. When the body starts moving, then the force of friction applies in the opposite direction to the direction of relative motion between the two surfaces. In general, we use the term "smooth" to refer to a frictionless interface and the term "rough" to refer to an interface with certain friction. We shall study more about friction in detail in a separate module.

The two contact forces i.e. normal and friction forces, are perpendicular to each other. The magnitude of net contact force, therefore, is given by the magnitude of the vector sum of two contact forces:

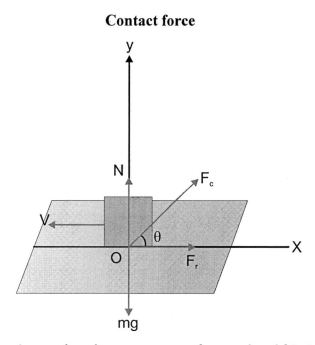

**Figure 4.15: The contact force is equal to the vector sum of normal and friction forces**

$$F_C = \sqrt{N^2 + F_f^2}$$

The tangent of the angle formed by the net contact force to the interface is given as,

Tan θ = N/F$_f$

# String Tension

String is an efficient medium to transfer force. We pull objects with the help of string from a convenient position. The string in taut condition transfers force as tension.

**Tension in string**

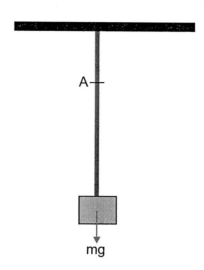

**Figure 4.16**

Let us consider a block hanging from the ceiling with the help of a string. In order to understand the transmission of force through the string, we consider a cross section at a point A as shown in the figure. The molecules across "A" attract each other to hold the string as a single piece.

**Tension in string**

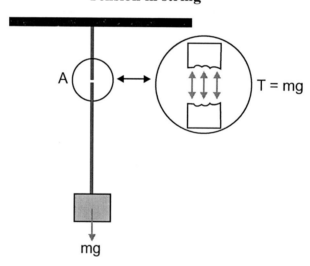

If we consider that the mass of the string is negligible, then the total downward pull is equal to the weight of the block (mg). The electromagnetic force at "A", therefore, should be equal to the weight acting in the downward direction. This is the situation at all points in the string and thus, the weight of the block is transmitted throughout the length of the string without any change in magnitude.

However, we must note that force is transmitted undiminished under three important conditions: the string is (i) taut (ii) inextensible and (ii) mass-less. Unless otherwise stated, these conditions are implied when we refer to string in the study of dynamics.

We must also appreciate that string is used with various combinations of pulleys to effectively change the direction of force without changing the magnitude. There is usually some doubt in one's mind about the direction of tension in the string. We see that directions of tension in the same piece of string are shown in opposite directions.

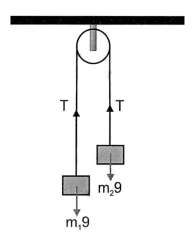

**Figure 4.17**

However, this is not a concern that should be overly emphasized. After all, the purpose of using string is to show that force is communicated undiminished (T), but with change in direction. We choose the appropriate direction of tension in relation to the body that is under focus for the study of motion. For example, the tension (T) is acting upward on the block, when we consider forces on the block. On the other hand, the tension is acting downward on the pulley, when we consider forces on the pulley. This inversion of direction of tension in a string is perfectly fine as tension works in opposite directions at any given intersection.

While considering string force as an element in the dynamic analysis, we should keep the following aspects in mind :

1:   If the string is taught and inextensible, then the velocity and acceleration of each point of the string (also of the objects attached to it) are the same.

**Inextensible string**

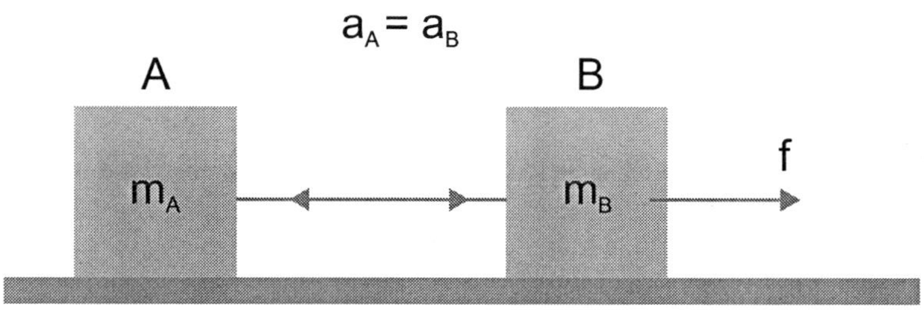

$$a_A = a_B$$

A                                          B

$m_A$                                 $m_B$        f

**Figure 4.18: The velocity and acceleration are the same at each point on the string.**

2:   If the string is "mass-less", then the tension in the string is the same at all points on the string.

**Mass-less string**

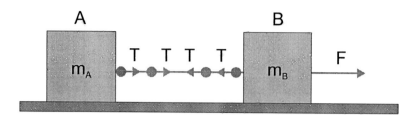

**Figure 4.19: The tension is the same at each point on the string**

3:    If the string has certain mass, then the tension in the string is different at different points. If the distribution of mass is uniform, we account for the mass of the string in terms of "mass per unit length ($\lambda$)" - also called the linear mass density of the string.

**String with certain mass**

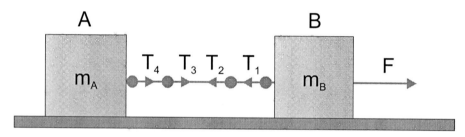

**Figure 4.20: The tension is different at different points on the string.**

4:    If "mass-less" string passes over a "mass-less" pulley, then the tension in the string is the same on the two sides of the pulley.

**Mass - less string and pulley**

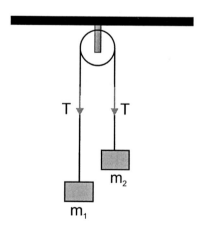

**Figure 4.21: The tensions in the string are the same on the two sides of the pulley.**

5:  If string, having certain mass, passes over a mass-less pulley, then the tension in the string is different on the two sides of the pulley.

**String and pulley**

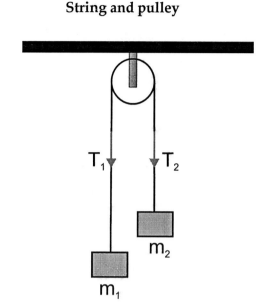

**Figure 4.22: The tensions in the string are different on the two sides of the pulley.**

6:  If "mass-less" string passes over a pulley with certain mass, and there is no slipping between string and pulley, then the tension in the string is different on the two sides of the pulley. Tensions of different magnitudes form the requisite torque required to rotate the pulley of certain mass.

# Spring

A spring is a metallic coil, which can be stretched or compressed. Every spring has a "natural length" that can be measured, while the spring is lying on a horizontal surface (shown at the top of figure below).

If we keep one end fixed and apply a force at the other end to extend it as shown in the figure, then the spring stretches by a certain amount, say $\Delta x$. In response to this, the spring applies an equal force in the opposite direction to resist deformation (shown in the figure at the middle).

# Spring force

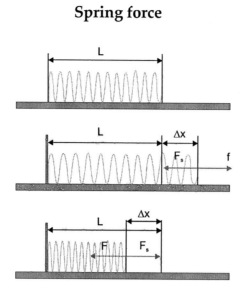

## Figure 4.23: Expansion and compression of spring

For a "mass-less" spring, it is

$$F_S \propto \Delta x$$

$$F_S = -k\Delta x$$

where "k" is called the spring constant, specific to a given spring. The negative sign is inserted to accommodate the fact that force exerted by the spring is opposite to the direction of change in the length of spring. This relation is known as Hooke's law and a spring, which follows Hooke's law, is said to be perfectly elastic.

Similarly, when an external force compresses the spring, it opposes compression in the same manner

The analysis of motion involves writing Newton's second law of motion in mutually perpendicular coordinate directions. For this, we carry out force analysis following certain simple steps. These steps may appear too many, but we become experienced while working with them. Some of the steps are actually merged. Nevertheless, we need to recognize the importance of each of the steps, as these are critical steps to ensure correct analysis of the motion / situation. The steps are:

1.  Identifying the body system

2.  Identifying external forces

3.  Identifying a suitable coordinate system

4.  Constructing a free-body diagram

5.  Resolving force along the coordinate axes

6.  Applying the second law of motion

The identification of the body system is based on the description of the problem. The body system decides the nature of force - whether it is internal or external. The selection of the body system should suit the requirement of the analysis i.e. the end objective of the analysis. This will become clear as we discuss this aspect subsequently in the module.

The identification of external forces means to identify forces which are applied on the object by other objects/sources, and exclude forces which the object applies on other objects. The important aspect of identification of external force is that external characterization depends on the definition of the body system. The same force, which is external to a body, may become an internal force for a different body system involving that body ( steps two and six are discussed elsewhere).

## Identifying body system

If we are studying a single or two body system, then there is no issue. However, let us consider an illustration here as shown in the figure. There are possibilities of having different combinations of bodies constituting different body systems.

**Identifying body system**

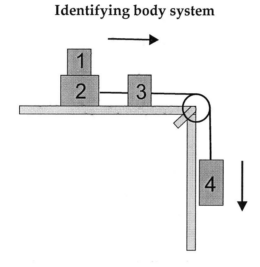

**Figure 4.24**

In selection of a body system, the guiding principle is to treat bodies as a single body system, when we know that bodies have common acceleration.

In the figure 4.24, the objects numbered 2 and 3 can be considered a single body system as two bodies have the same acceleration. Objects 3 and 4 cannot be combined as a single b ody system as they have acceleration of same magnitude, but with different direction. Objects 1 and 2 can be treated as a single body system, if two bodies do not have relative motion with respect to each other. Otherwise, we would need to treat them as separate body systems.

It is not always required that we must know the direction of acceleration beforehand. We can assume a direction with respect to one body system and then proceed to find the directions of the accelerations of the other body systems in the arrangement. Even if our assumption about the direction of acceleration is incorrect, the solution of the problem automatically corrects the direction. We shall see this aspect while working a specially designed example, highlighting the issue.

But the basic question is, why should we look for more than one body system in the first place? It is because such considerations will yield different sets of equations involving laws of motion and thus facilitate solution of the problem in hand.

## Identifying a suitable coordinate system

There is no rule in this regard. However, a suitable coordinate system facilitates easier analysis of the problem in hand. The guiding principle in this case is to select a coordinate system such that one of the axes aligns itself with the direction of acceleration and other is perpendicular to the direction of acceleration.

### Identifying a suitable coordinate system

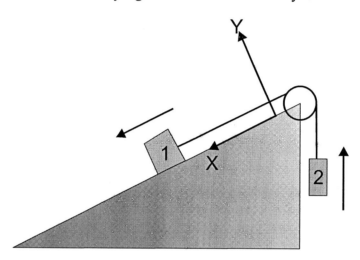

**Figure 4.25**

For the case as shown in the figure 4.25, the coordinate system for body 1 has its x-axis aligned with the incline, whereas its y-axis is perpendicular to the incline. In this case, when the bodies are in equilibrium, then we may align axes such that they minimize the requirement of taking components.

It must be realized here that we are free to utilize more than one coordinate system for a single problem. However, we would be required to appropriately choose the signs of vector quantities involved.

## Identifying a suitable coordinate system

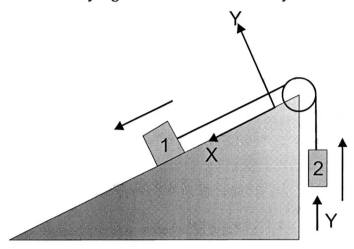

**Figure 4.26**

In the figure 4.26, we have selected two coordinate systems; one for body 1 and other for body 2. Note that the y-axis for body 2 is vertical, whereas the y–axis of body 1 is perpendicular to the incline. Such considerations are all valid as long as we maintain the sign requirements of the individual coordinate systems. In general, we combine "scalar" results from two different coordinate systems. For example, the analysis of force on body 2 yields the magnitude of tension in the string. This value can then be used for analysis of force on body 1, provided the string is mass-less and the tension in it transmits undiminished.

# Constructing a free-body diagram

A free body diagram is a symbolic diagram that represents the body system with a "point" and shows the external forces in both magnitude and direction. This is the basic free-body diagram, which can be supplemented with an appropriate coordinate system and some symbolic representation of acceleration. The free body diagram of body 1 is shown in the upper left corner of the figure:

## Constructing a free-body diagram

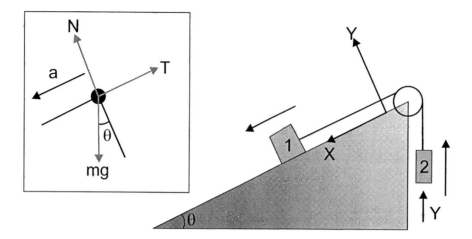

**Figure 4.27**

We may have as many free body diagrams as required for each of the body systems. If the direction of acceleration is known before hand, then we may show its magnitude and direction or we may assign an assumed direction of acceleration. In the latter case, the solution ultimately lets us decide whether the chosen direction was correct or not.

# What force system do we study?

We have discussed in the module titled "Newton's second law of motion" that we implicitly consider force the system on a body as concurrent forces. This is because Newton's second law of motion connects force (cause) with "linear" acceleration (effect). The word "linear" is implicit as there is no reference to angular quantities in the statement of the second law for translation.

We, however, know that such con-currency of forces can only be ensured if we consider point objects. What if we consider real three dimensional bodies? A three dimensional real body involves only translation, if external forces are concurrent (meeting at a common point) and coincides with the "center of mass" of the body. In this situation, the body can be said to be in pure translation. We need only one modification that we assign acceleration of the body to a specific point, called the "center of mass" (we shall elaborate on this aspect in separate module).

The real time situation presents real three dimensional bodies. We encounter situation in which forces are not concurrent. Still, we consider them to be concurrent (by shifting force in a parallel direction). It is because of the physical set up that constrains motion to be translational. For example, consider the motion of a block on an incline as shown in the figure.

**A block on an incline**

**(a) Friction induces a tendency to overturn.**

**(b) The weight counterbalances the tendency to overturn.**

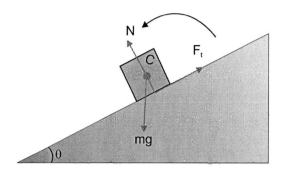

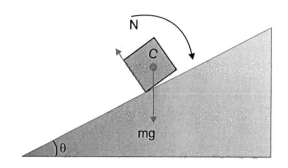

**Figure 4.28**

The forces weight (mg), normal force (N) and friction force (Ff) are not concurrent. The friction force, in fact, does not pass through the "center of mass" and as such induces a "turning" tendency. However, the block is not turned over, as restoring force due to the weight of the block is greater and inhibits turning of the block. In a nutshell, the block is constrained to translate without rotation. Here, translation is enforced not because forces are concurrent, but because the physical situation constrains the motion to be translational.

## Resolving force along the coordinate axes

The application of a force and resulting motion, in general, is three dimensional. It becomes convenient to analyze force (cause) and acceleration (effect) analysis along coordinate axes. This approach has the benefit that we get as many equations as there are axes involved. In turn, we are able to solve equations for as many unknowns.

We must know that the consideration in each direction is an independent consideration – not depending on the motion in other perpendicular directions. A force is represented by an equivalent system of components in mutually perpendicular coordinate directions.

The component of force vector along a direction (say x-axis) is obtained by :

**Component of force**

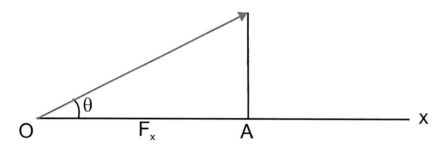

**Figure 4.29 : The angle is measured from the positive reference direction.**

$F_x = Fcos\theta$

where "$\theta$" is the angle that the force vector makes with the positive direction of the x-axis. For example, consider the force as shown in the figure. We need to find the component of a force along the x-axis. Here,

**Component of force**

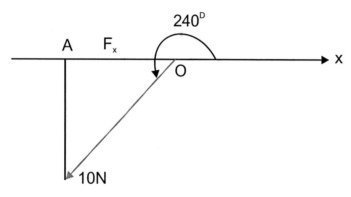

**Figure 4.30: 240 degree angle is considered.**

$F_x = Fcos\theta = 10cos240^0 = 10X-$

$\frac{1}{2} = -5\,N$

Alternatively, we consider only the acute angle that the force makes with the x-axis (we need not measure the reflex angle). The angle here is 60°. As such, the component of the force along the x-axis is:

## Component of force

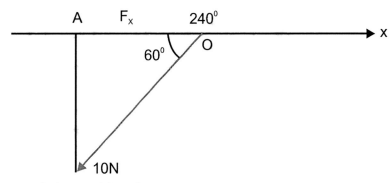

**Figure 4.31: 60 degree angle is considered.**

$F_x = F\cos\theta = 10\cos60^0 = 10X$

$\frac{1}{2} = 5\,N$

We decide the sign of the component by observing whether the projection of the force is in the direction of the x-axis or opposite to it. In this case, it is in the opposite direction. Hence, we apply the negative sign as,

$F_x = -F\cos\theta = -10\cos60^0 = -10X$

$\frac{1}{2} = -5\,N$

Clearly, if a force "$F$" makes angles $\alpha$, $\beta$ and $\gamma$ with the three mutually perpendicular axes, then :

## Components of force

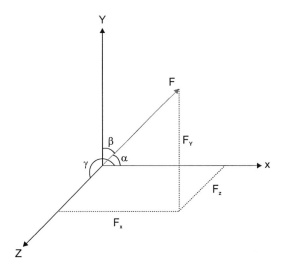

**Figure 4.32: The angle is measured from respective directions of axes.**

$F_x = F\cos\alpha$

$F_y = F\cos\beta$

$F_z = F\cos\gamma$

We can write the vector force using components as,

$F = F\cos\alpha i + F\cos\beta j + F\cos\gamma k$

In the case of a coplanar force, we may consider only one angle i.e. the angle that force makes with one of the coordinate directions. The other angle is a complement of this angle. If "θ" is the angle that force makes with the positive direction of the x-axis, then:

**Components of coplanar force**

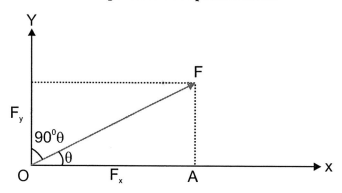

**Figure 4.33: Two components of a coplanar system**

$F_x = F\cos\theta$

$F_y = F\cos(90^0 - \theta) = F\sin\theta$

For further examples, please visit:

http://cnx.org/content/m14046/latest/

For questions and answers on Newton's Laws and its applications on pulleys, friction, and in accelerated frame of reference, please visit the website:

http://fountain.cnx.rice.edu:8280/content/m14041/latest/

http://fountain.cnx.rice.edu:8280/content/m14042/latest/

# PowerPoint Link:

Please refer to the end of the module lecture links.

# Discussion Question

Answer the following questions with references. Please remember to follow the standard APA referencing style.

For APA standards of references, please visit: http://owl.english.purdue.edu/owl/resource/560/01/

Also, respond in detail to one other post by fellow students.

4.1  Discuss the use of headrests on the seats in a car in terms of Newton's Laws. You need to explain all the terms used and provide a comprehensive answer.

4.2  Explain the validation of Newton's laws in an inertial framework by bringing examples of absolute and relativity movements

# Laboratory Activity and the link

Go to the link below and run the simulation:

http://phet.colorado.edu/simulations/sims.php?sim=Masses_and_Springs

Go to: http://phet.colorado.edu/en/contributions/view/2939

Open Spring Sim.doc

Answer all the questions.


<u>Now take the chapter 4 test (not included with this book)</u>

# Module 2 Lecture Links:

http://ocw.mit.edu/courses/physics/8-01-physics-i-classical-mechanics-fall-1999/video-lectures/lecture-6/

http://ocw.mit.edu/courses/physics/8-01-physics-i-classical-mechanics-fall-1999/video-lectures/lecture-7/

http://ocw.mit.edu/courses/physics/8-01-physics-i-classical-mechanics-fall-1999/video-lectures/lecture-8/

http://ocw.mit.edu/courses/physics/8-01-physics-i-classical-mechanics-fall-1999/video-lectures/lecture-10/

# Module 2 - Student's self-assessment

Please answer the following questions, which give you an indication of your standard of learning for this module:

- Could you describe Newton's three laws of motion?
- Could you apply Newton's laws to solve problems?
- Can you describe terminal velocity?
- What sections of the textbooks would you read for more information on the above?
- Did you enjoy your lesson?
- What aspect of the lesson was most interesting to you?
- Can you now confidently do the lesson's activities?

# Module 3
## CHAPTER-5

# Energy

## Objectives

At the end of this lesson, you should be able to:

1. Answer questions on potential and kinetic energy

2. Answer questions on combined PE, KE, and Force questions

## Lecture Notes

Work is a general term that we use in our daily life to assess execution or completion of a task. The basic idea is to define a quantity that can be used to determine both "effort" and "result". In physics also, the concept of work follows the same basic idea. However, it is completely "physical" in the sense that it recognizes only force as the "effort" and only displacement as the "result". There is no recognition of mental or any other effort that does not involve physical movement of a body.

For a constant force (F) applied on a particle, work is defined as the product of "component of force along the direction of displacement" and "the magnitude of displacement of the particle". Mathematically,

$W=(F\cos\theta)r=Fr\cos\theta$

where "$\theta$" is the angle between force and displacement vectors and "r" is the magnitude of displacement. The scalar component of force is also known as the projection of force. The SI unit of work is Newton - meter (N-m). This is equivalent to the unit of energy $kg-m^2s^{-2}$ i.e. Joule (J).

In order to appreciate the formula to compute work, let us consider an example. A block is being pulled by an external force "F" on a smooth horizontal plane as shown in the figure. The work (W) by force (F) is:

**A block pulled by external force**

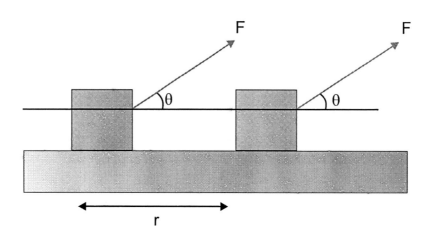

**Figure 5.1: External force "F" is a constant force**

$W = Fr\cos\theta$

The perpendicular forces i.e. normal force "N" and weight of block "mg" (not shown in the figure), do "no work" on the block as $\cos\theta = \cos 90° = 0$. An external force does the maximum work when it is applied in the direction of displacement. In that case, $\theta = 0°$, $\cos 0° = 1$ (maximum) and Work $W_G = Fr$.

# Work as vector dot product

Work involves two vector quantities - force and displacement - but work itself is a signed scalar quantity. Vector algebra provides the framework for such multiplication of vectors, yielding a scalar result via multiplication known as a dot product. The work done by the constant force as a dot product is:

$W = \boldsymbol{F.r} = Fr\cos\theta$

where "$\theta$" is the angle between force and displacement vectors.

# Computation of work

## Sign of work

Work is a signed scalar quantity. This means that it can be positive or negative, depending on the value of the angle between force and displacement. We shall discuss the significance of the sign of work in a separate module. However, we should be aware that the sign of work has specific meaning for the body on which force works. The sign determines the direction of energy exchange taking place between the body and its surrounding.

It is clear that the value of "cosθ" decides the sign of work. However, there is an easier method to determine the sign of work. We determine the magnitude of work considering projection of force and displacement - without any consideration of the sign. Once magnitude is calculated, we simply check whether the component of force and displacement are in same direction or in opposite directions. If they are in opposite directions, then we put a negative sign before the magnitude of work.

## Work by the named force

A body like the block on an incline is subjected to many forces viz: weight, friction, normal force and other external forces. Which of the forces do work? Is the work associated with net force, or any of the forces mentioned? In physics, we can relate work with any force or the net force working on the body. The only requirement is that we should mention the force involved. For this reason, we may be required to calculate work by any of the named forces.

It is, therefore, a good idea that we specify the force(s) that we have considered in the calculation of work. To appreciate this point, we consider a block being slowly raised vertically by hand to a height "h" as shown in the figure. As the block is not accelerated, the normal force applied by the hand is equal to the weight of the body:

**A block being raised slowly up**

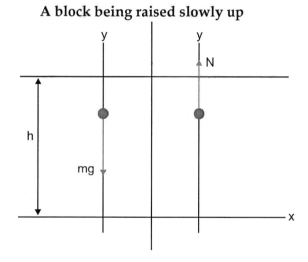

**Figure 5.2: Work done by normal force and gravity are different.**

$N = mg$

For gravity (gravitational force due to Earth), the force and displacement are opposite. Hence, work by gravity is negative.

The work done by gravity is,

$W_G = -(mg)h = -mgh$

For normal force applied by the hand on the block, force and displacement are both in the same direction. Hence, work done is positive:

The work by the hand is,

$W_H = NXh =$

$mgh$

Thus, the two instances of work are equal in magnitude, but opposite in sign. It can be easily inferred from the example here that work is positive, if both displacement and component of force along displacement are in the same direction; otherwise it is negative. It is also pertinent to mention that a subscripted notation for work as above is a good practice to convey the context of work. Finally, we should also note that net force on the body is zero. Hence, work by net force is zero - though works by individual forces are not zero.

# Examples

In the discussion above, we have made two points : (i) the sign of work can be evaluated either evaluating "cosθ" or by examining the relative directions of the component of force and displacement and (ii) work is designated to named force. Here, we select two examples to illustrate these points. The first example shows computation of work by friction - one of the forces acting on the body. The determination of the sign of the work is based on the evaluation of cosine of the angle between force and displacement. The second example shows the computation of work by gravity. The determination of the sign is based on the relative comparison of the directions of the component of force and displacement.

## Evaluation of cosine of angle

*Problem 1* : A block of 2 kg is brought up from the bottom to the top along a rough incline of length 10 m and height 5 m by applying an external force parallel to the surface. If the coefficient of kinetic friction between surfaces is 0.1, find work done by the friction during the motion. (consider, g = 10 $m/s^2$).

*Solution* : We see here that there are four forces on the block : (i) weight (ii) normal force (iii) friction and (iv) force, "F" parallel to incline. The magnitude of external force is not given. We are, however, required to find work by friction. Thus, we need to know the magnitude of friction and its direction. As the block moves up, kinetic friction acts downward. Here, displacement is equal to the length of incline, which is 10 m.

**Motion on a rough incline**

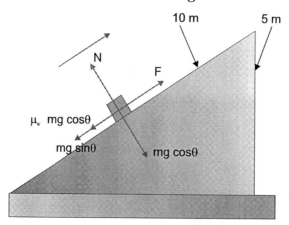

**Figure 5.3: The forces on the incline**

From the figure, it is clear that friction force is given as :

$F_k = \mu_k N = \mu_k mg cos\theta$

Here,

$$sin\theta = \frac{5}{10}$$

And $cos\theta = \sqrt{(1-sin^2\theta)}$

$$F = 0.1 \times 2 \times 10 \times \sqrt{\left(1 - \frac{5^2}{10^2}\right)}$$
$F = 2 \times 0.866 = 1.732 N$

To evaluate work in terms of "Frcosφ", we need to know the angle between force and displacement. In this case, this angle is 180° as shown in the figure below.

Note:

We denote "φ" instead of "θ" as angle between force and displacement to distinguish this angle from the angle of incline.

## Motion on a rough incline

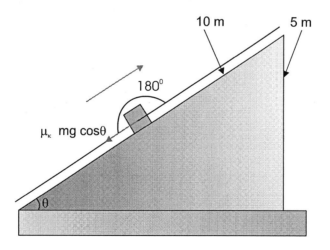

**Figure 5.4: The angle between friction and displacement**

$W = Frcos\varphi$

$W_F = Frcos\varphi = 1.732 \times 10 \times cos180^0$

$W_F = 1.732 \times 10 \times (-1) = -17.32 J$

This example brings out the concept of work by named force (friction). The important point to note here is that we could calculate work by friction even though we did not know the magnitude of force of external force, "F". Yet another point to note here is that computation of work by friction is actually independent of - whether the block is accelerated or not. In addition, this example illustrates how the evaluation of the cosine of the angle between force and displacement determines the sign of the work.

### Relative comparison of directions

*Problem 2:* A block of 2 kg is brought up from the bottom to the top along a smooth incline of length 10 m and height 5 m by applying an external force parallel to the surface. Find work done by the gravity during the motion. (consider, g = 10 *m/s²* ).

*Solution:* In this problem, the incline is smooth. Hence, there is no friction at the contact surface. Now, the component of gravity (weight) along the direction of the displacement is:

**Motion on a smooth incline**

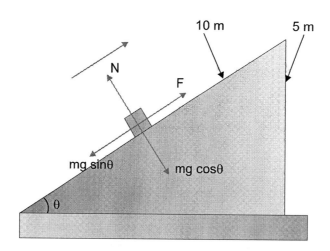

**Figure 5.5: Forces on the block (except external force)**

$Mg\sin \theta = 2 \times 10 \times \frac{5}{10} = 10N$

We, now, determine the magnitude of work without taking into consideration of the sign of the work. The magnitude of work by gravity is:

$W_G = mg\sin\theta \times r = 10 \times 10 = 100J$

Once magnitude is calculated, we compare the directions of the component of force and displacement. We note here that the component of weight is in the opposite direction to the displacement. Hence, work by gravity is negative. As such, we put a negative sign before magnitude.

$W_g = -100J$

It is clear that this second method of computation is the easier of the two approaches. One of the simplifying aspects is that we need to calculate cosine of only the acute angle to determine the magnitude of work without any concern about directions. We may then assign sign, subsequent to calculation of the magnitude of work.

# Work in three dimensions

We can extend the concept of work to motion in three dimensions. Let us consider three dimensional vector expressions of force and displacement:

$$F = F_x\mathbf{i} + F_y\mathbf{j} + F_z\mathbf{k}$$

and

$$R = x\mathbf{i} + y\mathbf{j} + z\mathbf{k}$$

The work as a dot product of two vectors is:

$$W = F.r = (F_x\mathbf{i}+F_y\mathbf{j}+F_z\mathbf{k}).(x\mathbf{i}+y\mathbf{j}+z\mathbf{k})$$

$$W = F_x x + F_y y + F_z z$$

From the point of view of computing work, we can calculate "work" as the sum of the products of scalar components of force and displacement in three mutually perpendicular directions along the axes with appropriate sign. Since respective components of force and displacement are along the same direction, we can determine work in each direction with appropriate sign. Finally, we compute their algebraic sum to determine work by force, "F".

# Work by a variable force

We have defined work for constant force. This condition of constant force is, however, not a limitation as we can use calculus to compute work by a variable force. In order to keep the derivation simple, we shall consider force and displacement along the same straight line or direction. We have already seen that the calculation of work in a three dimensional case is equivalent to calculation of work in three mutually perpendicular directions.

A variable force can be approximated to be a series of constant forces of different magnitude as applied to the particle. Let us consider that force and displacement are in the same x-direction.

work by variable force

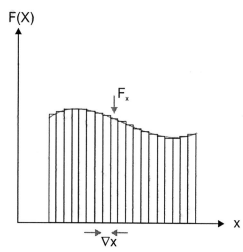

**Figure 5.6: Work is given by the area under the plot.**

For a given small displacement ($\Delta$x), let $F_x$ be the constant force. Then, the small amount of work for covering a small displacement is:

$$\Delta W = F_x \Delta x$$

We note that this is the area of the small strip as shown in the figure above. The work by the variable force over a given displacement is equal to the sum of all such small strips,

$$W = \sum \Delta W = \sum F_x \Delta x$$

For better approximation of the work by the variable force, the strip is made thinner as $\Delta$x-->0, whereas the number of strips tends to be infinity. For the limit,

$$W = \lim_{\Delta x \to 0} \sum F_x \Delta x$$

This limit is equal to the area of the plot defined by the integral of force function F(x) between two limits,

$$W = \int_{x_1}^{x_2} F(x) dx$$

# Example

*Problem 3*: A particle moves from point A to B along the x - axis of a coordinate system. The force on the particle during the motion varies with displacement in the x-direction as shown in the figure. Find the work done by the force.

## Work by variable force

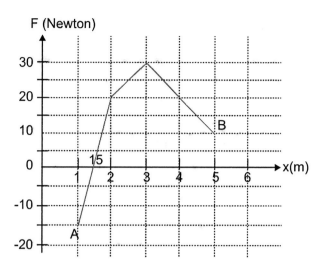

**Figure 5.7: Force - displacement plot.**

*Solution* : The work done by the force is:

$$W = \int_{x_1}^{x_2} F(x)dx = \text{Area between plot and x-axis within the limits}$$

Now, the area is:

$$W = -\frac{1}{2} X \, x0.5x15 + \frac{1}{2} x0.5x20x1 + \frac{1}{2}(30 + 20) + \frac{1}{2} X \, 2 \, x \, (30 + 10)$$

$$W = -3.75 + 5 + 25 + 40 = 66.25 J$$

# Energy

Energy is one of two basic quantities (besides mass) that constitute our universe. Because of the basic nature and generality involved with energy, it is difficult to propose an explicit definition of energy which is meaningful in all situations. We have two options. Either (i) we define energy a bit vaguely for all situations or (ii) we define energy explicitly in limited context. For all situations, we can say that energy is a quantity that measures the "state of matter".

There are wide varities or forms of energy. The meaning of energy in a particular context is definitive. For example, kinetic energy, which is associated with the motion (speed) of a particle, has a mathematical expression to compute it precisely. Similarly, energy has concise and explicit meaning in electrical, thermal, chemical and such specific contexts.

We need to clarify here that our study of energy and related concepts presently deals with particle or particle-like objects such that particles composing the object have the same motion. We shall extend these concepts subsequently to a group of particles and situations where particles composing an object may have different motions (rotational motion).

For easy visualization, we (in a mechanical context) relate energy with the capacity of a body to do work ("work" as defined in physics). This definition enables us to have the intuitive appreciation of the concept of energy. The capacity for doing work, here, does not denote that the particle will do the amount of work as indicated by the level of energy. For example, our hand has the capacity to do work. However, we may end up doing "no work", like when pushing a building with our hand.

On the other hand, thermal energy (non-mechanical context) of a body cannot be completely realized as work. This is actually one of the laws of thermodynamics. Hence, we may appreciate the connection between energy and work, but should avoid defining energy in terms of the capacity to do work. We shall soon find that work is actually a form of energy, which is in "transit" between different types of energies.

# Kinetic energy

One of the most common types of energy that we come across in our day to day life is the energy of motion. This energy is known as kinetic energy and defined for a particle of mass "m" and speed "v" as:

$$K = \frac{1}{2}mv^2$$

Kinetic energy arises due to "movement" of a particle. The main characteristics of kinetic energy are as follows:

- The expression of kinetic energy involves the scalar quantities mass "m" and speed "v". Importantly, it involves speed i.e. the magnitude of velocity - not vector velocity. Therefore, kinetic energy is a scalar quantity.

- Both mass "m" and speed "v" are positive scalar quantities. Therefore, kinetic energy is a positive scalar quantity. This means that a particle cannot have negative kinetic energy.

- Kinetic energy of a particle, at rest, is zero.

- The greater the speed or mass, the greater is the kinetic energy and vice-versa.

The SI unit of kinetic energy is $kg-m^2s^{-2}$, which is known as "Joule". Since kinetic energy is a form of energy, the Joule (J) is the SI unit of all types of energy.

# Work and kinetic energy

We have reached the point where we can attempt to relate work with energy (actually kinetic energy). The relationship is easy to visualize in terms of the motion of a body.

In order to fully appreciate the connection between work and kinetic energy, we consider an example. A force is applied on a block such that the component of force is in the direction of the displacement as shown in the figure below. Here, work by force on the block is positive. During the time force does positive work, the speed and consequently kinetic energy of the block increases ( $K_f > K_i$ ) as the block moves ahead with certain acceleration. We shall know later that the kinetic energy of the block increases by the amount of work done by net external force on it. This is what is known as the "work - kinetic energy" theorem and will be the subject matter of a separate module.

**Work done on a block**

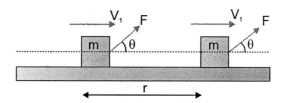

Figure 5.8: **The component of external force and displacement are in the same direction.**

For the time being, we concentrate on work and determine its relationship with energy in qualitative terms. Since kinetic increases by the amount of work done on the particle, it follows that work, itself, is an energy which can be added as kinetic energy to the block. The other qualification of the work is that it is the "energy" which is transferred by the force from the surrounding area to the block. Clearly, force here is an agent, which does the work to increase the speed of the object and hence to increase its kinetic energy. In this sense, "work" by a force is the energy transferred "to" the block, on which force is applied.

We, now, consider the reverse situation as illustrated in the figure below. A force is applied to retard the motion of a block. Here, the component of force is in the opposite direction to the displacement. The work done on the block is negative. The kinetic energy of the block decreases ( $K_f < K_i$ ) by the amount of work done by the force. In this case, kinetic energy is transferred "out" of the block and is equal to the amount of negative work by force. Here, "work" by a force is the energy transferred "from" the block, on which force is applied. Thus, we can define "work" as,

**Work done on a block**

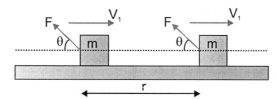

Figure 5.9: **The component of external force and displacement are in opposite directions.**

## Definition 1: Work

Work is the energy transferred by the force "to" or "from" the particle on which force is applied.

It is also clear that a positive work means transfer of energy "to" the particle and negative work means transfer of energy "from" the particle. Further, the term "work done" represents the process of transferring energy to the particle by the force.

# Work done by a system of forces

The connection of work with kinetic energy is true for a special condition. The change in kinetic energy resulting from doing work on the body refers to work by net force - not work by any of the forces operating on the body. This fact should always be kept in mind, while attempting to explain motion in terms of work and energy.

Above observation is quite easy to comprehend. Consider motion of a block on a rough incline. As the block comes down the incline, the component of gravity does the positive work i.e. transfers energy from the gravitational system to the block. As a consequence, the speed and therefore the kinetic energy of the block increases. On the other hand, friction acts opposite to displacement and hence does negative work. It draws some of the kinetic energy of the block and transfers the same to the surrounding area as heat. We can see that gravity increases speed, whereas friction decreases speed. Since work by gravity is more than work by friction in this case, the block comes down with increasing speed. The point is that net change in speed and hence kinetic energy, is determined by both the forces operating on the block. Hence, we should think of the relation between work by "net" force and kinetic energy.

As mentioned earlier, a body under consideration may be subjected to more than one force. In that case, if we have to find the work by the net force, then we can adopt either of two approaches:

(i)  Determine the net (resultant) force. Then, compute the work by net force.

$$F = \sum F_i$$

$$W = F \cdot r = Fr\cos\theta$$

(ii)  Compute work by individual forces. Then, sum the works to compute work by net force.

$$W_i = F_i \cdot r_i$$

$$W = \sum W_i$$

Either of the two methods yields the same result. However, there is an important aspect about the procedures involved. If we determine the net force first, then we shall require the use of a free body diagram and a coordinate system to analyze forces to determine net force. Finally, we use the formula to compute work by the net force. On the other hand, if we follow the second approach, then we need to find the product of the components of individual forces along the displacement. Finally, we carry out the algebraic sum to find the work by the net force.

This is a very significant point. We can appreciate this point fully, when the complete framework of the analysis of motion, using concepts of work and energy, is presented. For the time being, we should understand that work together with energy provides an alternative to analyzing motion. Imagine if we could completely do away with the "free body diagram" and "coordinate system". Indeed, it is a great improvisation and simplification, if we can say so. However, this simplification comes at the cost of a more involved understanding of the physical process. The meaning of this paragraph will be clearer as we go through the modules on the related topics. For the time being, let us work out a simple example illustrating the point.

## Example

*Problem 1:* A block of mass "10 kg" slides down through a length of 10 m over an incline of 30°. If the coefficient of kinetic friction is 0.5, then find the work done by the net force on the block.

*Solution* : There are two forces working on the block along the direction of displacement (i) the component of gravity and (ii) the kinetic friction in the direction opposite displacement.

**A block on a rough incline**

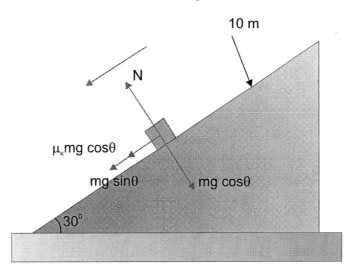

**Figure 5.10: Work by gravity is positive and work by friction is negative.**

The component of gravity along displacement is "mg sinθ" and is in the direction of displacement. Work by gravity is:

$$W_G = mgL\sin 30° = 10 \times 10 \times 10 \times \frac{1}{2} = 500 J$$

Friction force is " $\mu_K mg \sin\theta$ " and is in the opposite direction to that of displacement.

$$W_F = -\mu_K mgL\cos 30° = -0.5 \times 10 \times 10 \times 10 \times \frac{\sqrt{3}}{2} = -433 J$$

Hence, work by net force is:

$$W_{net} = W_G + W_F = 1000 - 433 = 567 J$$

The important thing to note here is that the reference of the component of force is with respect to displacement - not with respect to any coordinate direction. This is how we eliminate the requirement of a coordinate system to calculate work.

Imagine if we first find the net force. It would be tedious as friction acts along the incline, but gravity acts vertically at an angle with the incline. Even if we find the net force, its angle with displacement would be required to evaluate the expression of work. Clearly, calculation of work for individual force is easier. However, we must keep in mind that we can employ this technique only if we know the forces beforehand.

## Analysis of motion

The concepts of work and energy together are used to analyze motion. The basic idea is to analyze motion such that it does not require intermediate details of the motion like velocity, acceleration and path of motion. The independence from the intermediate details is the central idea that makes work - energy analysis so attractive and elegant. It allows us to analyze circular motion in a vertical plane, motion along paths which are not straight line, and a host of other motions, which cannot be easily analyzed with the laws of motion. This is possible because we find that work by a certain class of force is independent of path. Further, under certain situations, the analysis is independent of details of attributes like velocity and acceleration.

## Path of motion

In this section, let us examine the issue of path independence. Does the work depend on the path of motion? The answer is both yes and no. Even though computation of work involves displacement - a measurement in terms of end points - work is not always free of the path involved. The freedom of path depends on the nature of force. We shall see that work is independent of path for forces like gravity. Work only depends on the vertical displacement and is independent of horizontal displacement. This is so due to the horizontal component of Earth's gravity is zero. However, work by force like friction depends on how long (distance) a particle moves on the actual path.

The class of force for which work is independent of path is called "conservative" force; others are called "non-conservative". Therefore, we can say that work is independent of path for conservative force and is dependent for non-conservative force. This is the subject matter of a separate module on "conservative force" and as such, we shall not elaborate on the concept further.

This limited independence may appear to be disadvantageous. A complete independence for all forces would have allowed us to analyze motion without intermediate details in all situations. However, an important point here is that major forces in nature are conservative forces - gravitational and electromagnetic forces. This allows us to devise techniques to calculate work by "non-conservative force" indirectly without details, using other concepts (work - kinetic energy theory) that we are going to develop in the next module.

## Details of motion

Let us analyze the details of motion. Does work depend on whether a body is moved with acceleration or without acceleration? We can have a look at the illustration, in which we raise a block slowly against gravity through a certain height. Here, net force on the block is zero. Hence, work by net force is zero.

**Work in raising a body**

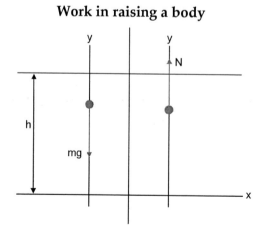

**Figure 5.11: Work by net force is zero.**

Now, let us modify the illustration a bit. We raise the block with a certain constant velocity to the same height. As velocity is constant, it means that there is no acceleration and the net force on the block is zero. Hence, work by net force is zero. However, if we raise the block with some acceleration, would it affect the amount of work by net force? The presence of acceleration means net force and as such, work by net force is not zero. Greater net force will mean greater work and acceleration.

In order to get the picture, we now look at the illustration from a different perspective. What about the work by component forces, like gravity or normal force, as applied by the hand? Work by individual force is the multiplication of force and the displacement. Since gravity is constant near the surface, work by gravity is constant for the given displacement. As a matter of fact, work by gravity is only dependent on vertical displacement.

When the block is raised slowly or with constant velocity, the net force is zero. In these circumstances, the work by normal force is equal to the work by gravity. This is an important deduction. It allows us to compute work by either force without any reference to velocity and acceleration.

We see that the work by conservative force is independent of the details of motion. Under certain situations, when work by net force is zero, we can determine work by other force(s) in terms of work by the conservative force. Thus, the basic idea is to make use of the features of conservative force to simplify analysis such that our consideration is independent of the details of motion.

## Role of Newton's laws of motion

The discussion so far leads us to identify distinguishing features of laws of motion on one hand and work-energy concepts on the other - as far as the analysis of motion is concerned.

The main distinguishing feature of the application of the laws of motion is that we should know the details of motion to analyze the same. This feature is both a strength and a weakness. It is a strength in the sense that if we have to know the details like acceleration, then we are required to analyze motion in terms of the laws of motion. "Work - energy" does not provide details.

On the other hand, if we have to know the broad parameters like energy or work, then "Work - energy" provides the most elegant solution. In fact, we would find that analysis by "Work - energy" is often supplemented with analysis by laws of motion to obtain detailed results.

Clearly, when both frameworks are used in tandem, we get the best of both worlds. The example here highlights this aspect. Carefully, note how two concepts are combined to achieve the result.

For work and energy applications, please visit: http://cnx.org/content/m14110/latest/

Work is itself energy, but plays a specific role with respect to other forms for energy. Its relationship with different energy forms will automatically come to the fore as we investigate them. In this module, we shall investigate the relationship between work and kinetic energy.

To appreciate the connection between work and kinetic energy, let us consider a block, which is moving with a speed "v" in a straight line on a rough horizontal plane. The kinetic friction opposes the motion and eventually brings the block to rest after a displacement, say "r".

**A block is brought to rest by friction**

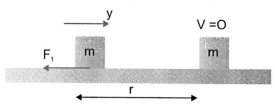

**Figure 5.12: Friction applies in opposite direction to displacement.**

Here, kinetic friction is equal to the product of coefficient of kinetic friction and normal force applied by the horizontal surface on the block,

$$F_k = \mu_k N = \mu_k mg$$

Kinetic friction opposes the motion of the block with deceleration, a:

$$a = \frac{Fk}{m} = \frac{\mu kmg}{m}$$

$$a = \mu_k g$$

Considering motion in the x-direction and using the equation of motion for deceleration, $v_2{}^2 = v_1{}^2 - 2ar$ , we have:

$$0 = v^2 - 2ar$$

$$v^2 = 2ar = 2\mu_k gr$$

Thus, kinetic energy of the block in the beginning of motion is:

$$K = \frac{1}{2}m\, v^2 = \frac{1}{2} X\, m^2 \mu kgr = \mu kmgr$$

A close inspection of the expression of initial kinetic energy as calculated above reveals that the expression is equal to the magnitude of work done by the kinetic friction to bring the block to rest from its initial sate of motion. The magnitude of work done by the kinetic friction is:

$$W_F = F_k r = \mu_k mgr = K$$

This brings us to a new definition of kinetic energy:

*Definition 1:* **Kinetic energy**

Kinetic energy of a particle in motion is equal to the amount of work done by an external force to bring the particle to rest.

# Work - kinetic energy theorem

The work - kinetic energy theorem is a generalized description of motion - not specific to any force type like gravity or friction. We shall, here, formally write the work - kinetic energy theorem considering an external force. The application of a constant external force results in the change in kinetic energy of the particle. For the time being, we consider a "constant" external force. At the end of this module, we shall extend the concept to variable force as well.

Let $v_i$ be the initial speed of the particle, when we start observing motion. Now, the acceleration of the particle is:

**A force moves the block on a horizontal surface**

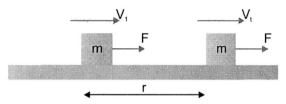

**Figure 5.13: Force does work on the block.**

$$a = \frac{F}{m}$$

Let the final velocity of the particle be $v_f$. Then using equation of motion, $v_f^2 = v_i^2 + 2ar$,

$$v_f^2 - v_i^2 = 2(F/m)\ r$$

Multiplying each term by 1/2 m, we have:

$$\text{Fr} = K_f - K_i = W$$

This equation is known as the work - kinetic energy theorem. In other words, change in kinetic energy resulting from the application of external force(s) is equal to the work done by the force(s). Equivalently, work done by the force(s) in displacing a particle is equal to the change in the kinetic energy of the particle. The above work - kinetic energy equation can be rearranged as:

$$K_f = K_i + W$$

In this form, the work - kinetic energy theorem states that kinetic energy changes by the amount of work done on the particle. We know that work can be either positive or negative. Hence, positive work results in an increase of the kinetic energy and negative work results in a decrease of the kinetic energy by the amount of work done on the particle. It is emphasized here for clarity that "work" in the theorem refers to work by "net" force - not individual force.

## Work - kinetic energy theorem with multiple forces

Extension of the work - kinetic energy theorem to multiple forces is simple. We can determine the net force of all external forces acting on the particle, compute work by the net force, and then apply ther work - kinetic energy theorem. This approach requires that we consider the free body diagram of the particle in the context of a coordinate system to find the net force on it.

$$F_N = F_1 + F_2 + F_3 + \ldots \ldots + F_n$$

The work - kinetic energy theorem is written for the net force as:

$$K_f - K_i = F_N.r$$

Where $F_N$ is the net force on the particle. Alternatively, we can determine work done by individual forces for the displacement involved and then sum them to equate with the change in kinetic energy. Most favor this second approach as it does not involve vector consideration with a coordinate system.

$$K_f - K_i = \sum W_i$$

# Application of Work - kinetic energy theorem

The work - kinetic energy theorem is not an alternative to other techniques available for analyzing motion. What we want to emphasize here is that it provides a specific technique to analyze motion, including situations where details of motion are not available. The analysis typically does not involve intermediate details in certain circumstances. One such instance is illustrated here.

In order to illustrate the application of the "work - kinetic energy" theory, we shall work with an example of a block being raised along an incline. We do not have information about the nature of motion - whether it is raised along the incline slowly or with constant speed or with varying speed. We also do not know whether the applied external force was constant or varying. However, we know the end conditions: the block was stationary at the beginning of the motion and at the end of motion. So,

$$K_f - K_i = 0$$

This means that work done by the forces on the block should sum up to zero (according to the "work - kinetic energy" theorem). If we know all other forces but one "unknown" force, and also work done by the known forces, then we are in position to know the work done by the "unknown" force.

We should know that application of the work-kinetic energy theorem is not limited to cases where the initial and final velocities are zero or equal, but can be applied also to situations where velocities are not equal. We shall discuss these applications with references to specific forces like gravity and spring force in separate modules.

It is also inferred from the discussion above that the other class of forces known as "non-conservative" force do not meet the requirement as needed for being conservative. They transfer energy "from" the object in motion just like conservative force, but they do not transfer this energy "to" the potential energy of the system to regain it during reverse motion. Instead, they transfer the energy to the system in an energy form which cannot be used by the force to transfer energy back to the object in motion. Friction is one such non-conservative force.

Let us, now, consider the motion of a block projected on a rough incline instead of a smooth incline. Here, force due to gravity and friction act on the block. During the upward motion, force due to gravity and friction together oppose motion and do negative work on the block. As the block is subjected to greater force than earlier, the block travels a lesser distance (L'). Work done by force due to gravity and friction duringthe upward motion are :

## A block projected on a rough incline plane

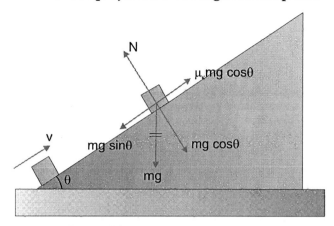

**Figure 5.14: The force due to gravity and friction do work on the block.**

$W_{G(up)} = -mgL'\sin\theta$

$W_{F(up)} = -\mu_k mgL'\sin\theta$

Total work done during upward motion is:

$W_{up} = -mgL'(\sin\theta + \mu_k\cos\theta)$

On the other hand, force due to gravity does positive work, whereas friction again does negative work during downward motion. Work done by force due to gravity and friction during downward motion are,

$W_{G(down)} = mgL'\sin\theta$

$W_{F(down)} = -\mu_k mgL'\sin\theta$

Total work done during downward motion is:

$W_{down} = mgL'(\sin\theta - \mu_k\cos\theta)$

Total work during the motion along a closed path is,

$W_{total} = -2mgL'\mu_k\cos\theta$

Here, network in the closed path motion is not zero as in the case of motion with conservative force. In fact, network in a closed path motion is equal to work done by the non-conservative force. We observe that friction transfers energy from the object in motion to the "Earth - Incline - block" system in the form of thermal energy - not as potential energy. Thermal energy is associated with the motion of atoms/ molecules composing the block and incline. Friction is not able to transfer energy "from" thermal energy of the system "to" the object, when motion is reversed in a downward direction. In other words, the energy withdrawn from the motion is not available for reuse by the object during its reverse motion.

We summarize important points about the motion, which is interacted by conservative force:

- The speed and kinetic energy of the object on return to initial position are lesser than initial values in a closed path motion.

- Non - conservative force does not transfer energy "from" the system "to" the object in motion.

- Non - conservative force transfers energy between the kinetic energy of the object in motion and the system via energy forms other than potential energy.

- Total work done by non-conservative force in a closed path motion is not zero.

## Path independence of conservative force

We have already seen that work done by gravity in a closed path motion is zero. We can extend this observation to other conservative force systems as well. In general, let us conceptualize what we have learned about conservative force. We imagine a closed path motion. We imagine this closed path motion to be divided in to two motions between points "A" and "B". Starting from point "A" to point "B" and then ending at point "A" via two work paths named "1" and "2" as shown in the figure. As observed earlier, the total work by the conservative force for the round trip is zero.

**Motion along closed path**

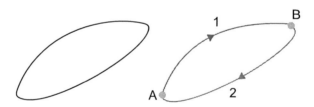

**Figure 5.15**

$W=W_{AB1}+W_{BA2}=0$

Let us now change the path for motion from A to B by another path, shown as path "3". Again, the total work by the conservative force for the round trip via new route is zero:

**Motion along closed path**

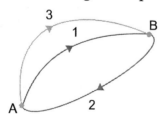

**Figure 5.16**

$W = W_{AB3} + W_{BA2} = 0$

Comparing two equations,

$W_{AB1} = W_{AB3}$

Similarly, we can say that work done for motion from A to B by conservative force along any of the three paths are equal:

$W_{AB1} = W_{AB2} = W_{AB3}$

We summarize the discussion as:

1:  Work done by conservative force in any closed path motion is zero. The word "any" is important. This means that the configuration of path can be shortest, small, large, straight, two dimensional, three dimensional etc. There is no restriction about the path of motion so long only conservative force(s) are the ones interacting with the object in motion.

**Motion along closed path**

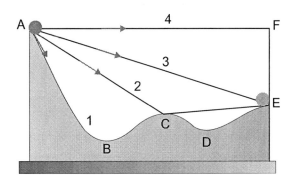

**Figure 5.17**

2:  Work done by conservative force(s) is independent of the path between any two points. This has a great simplifying implication in analyzing motions, which otherwise would have been tedious at the least. Four paths between "A" and "E" as shown in the figure are equivalent in the context of work done by conservative force. We can select the easiest path for calculating work done by the conservative force(s).

3:  The system with conservative(s) force provides a mechanical system, where energy is made available for reuse and where energy does not become unusable for the motion.

For the application of the work-kinetic energy theorem, work by gravity, spring force, and conversation of force, please visit:

http://cnx.org/content/m14104/latest/

# PowerPoint Link:

Please refer to the end of the module lecture link.

# Discussion Question

Answer the following questions with references. Please remember to follow the standard APA referencing style.

For APA standards of references, please visit: http://owl.english.purdue.edu/owl/resource/560/01/

Also, respond in detail to one other post by fellow students.

**5.1** The great Roman Poet, Lucretius wrote, " Things cannot be born from nothing, cannot when begotten brought back to nothing." This statement became an accepted theory as a result of the work of the chemist, Lavoisier in the 18th century. This is the law of conservation of matter. Similarly, the law of conservation of energy is an important law in physics. Can you discuss the validity of the conservation of energy in terms of potential (gravitational or elastic) and kinetic energy by giving real-life applications?

**5.2** Work done = Average Force X distance – Explain this equation ( using calculus) with examples.

# Laboratory Activity and the Link

Go to the link below and run the simulation:

http://phet.colorado.edu/simulations/sims.php?sim=The_Ramp

Go to: http://phet.colorado.edu/teacher_ideas/view-contribution.php?contribution_id=605

Open The Ramp PhET Lab.doc

Answer all the questions.

**Now take the chapter 5 test (not included with this book)**

# Module 3 Lecture Link:

http://ocw.mit.edu/courses/physics/8-01-physics-i-classical-mechanics-fall-1999/video-lectures/lecture-11/

# Module 3 – Student's self-assessment

Please answer the following questions, which give you an indication of your standard of learning for this module:

- Could you describe the potential energy?
- Could you describe kinetic energy?
- Can you apply knowledge to solve problems involving KE and PE and force?
- What sections of the textbook would you read for more information on the above?
- Did you enjoy your lesson?
- What aspect of the lesson was most interesting to you?
- Can you now confidently do the lesson's activities?

# Module 4

## CHAPTER-6

# Momentum and Collisions

## Objectives

At the end of this lesson, you should be able to:

1. Answer questions on momentum, impulse, and changes

2. Answer questions on elastic and inelastic collisions.

## Lecture Notes

We have briefly defined linear momentum, while describing Newton's second law of motion. The law defines force as the time rate of linear momentum of a particle. It directly provides a measurable basis for the measurement of force in terms of mass and acceleration of a single particle. As such, the concept of linear momentum is not elaborated on or emphasized for a single particle. However, we shall see in this module that linear momentum becomes a convenient tool to analyze the motion of a system of particles, particularly with reference to internal forces acting inside the system.

It will soon emerge that Newton's second law of motion is more suited for the analysis of the motion of a particle like objects, whereas the concept of linear momentum is more suited when we deal with the dynamics of a system of particles. Nevertheless, we must understand that these two approaches are interlinked and equivalent. Preference to a particular approach is basically a question of the suitability of the analysis of the situation.

Let us now review the main points of linear momentum as described earlier :

*i.* It is defined for a particle as a vector in terms of the product of mass and velocity.

$$p=mv$$

The small "$p$" is used to denote linear momentum of a particle and capital "$P$" is used for linear momentum of the system of particles. Further, these symbols distinguish linear momentum from angular momentum ($L$) as applicable in the case of rotational motion. By convention, a simple reference to "momentum" means "linear momentum".

*ii.* Since mass is a positive scalar quantity, the directions of linear momentum and velocity are the same.

*iii.* In a physical sense, linear momentum is said to signify the "quantum or quantity of motion". This is because a particle with higher momentum generates greater impact, when stopped.

*iv.* The first differentiation of linear momentum with respect to time is equal to the external force on the single particle.

$$F_{Ext.} = \frac{dp}{dt} = ma$$

## Momentum of a system of particles

The concept of linear momentum for a particle is extended to a system of particles by summing the momentum of individual particles. However, this sum is a vector sum of momentums. We need to either employ vector addition or equivalent component summation with appropriate sign convention as discussed earlier. Linear momentum of a system of particles is, thus, defined a :

### *Definition 1: Momentum of a system of particles*

The linear momentum of a system of particles is the vector sum of linear momentums of individual particles.

**Momentum of a system of particles**

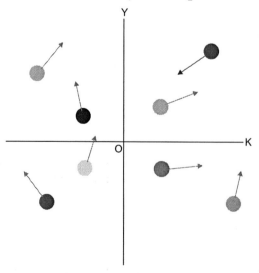

**Figure 6.1: Particles moving with diferent velocities.**

$$p = m_1 v_1 + m_2 v_2 + \ldots\ldots\ldots + m_n v_n$$

$$p = \sum m_i v_i$$

From the concept of "center of mass", we know that:

$$M v_{COM} = m_1 v_1 + m_2 v_2 + \ldots\ldots\ldots + m_n v_n$$

Comparing two equations,

$$P = M v_{COM}$$

The linear momentum of a system of momentum is, therefore, equal to the product of total mass and the velocity of the COM of the system of particles.

## External force in terms of momentum of the system

Just like the case for a single particle, the first differentiation of the total linear momentum gives the external force on the system of particles :

$$F_{Ext.} = \frac{dP}{dt}$$

$$F_{Ext} = M a_{COM}$$

This is the same result that we had obtained using the concept of center of mass (COM) of the system of particles. The application of the concept of linear momentum to a system of particles, however, is useful in the expanded form, which reveals the important aspects of "internal" and "external" forces :

$$F_{Ext.} = m_1 a_1 + m_2 a_2 + \ldots\ldots\ldots + m_n a_n$$

The right hand expression represents the vector sum of all forces on individual particles of the system.

$$F_{Ext.} = F_1 + F_2 + \ldots\ldots\ldots + F_n$$

This relation is slightly ambiguous. The left hand side symbol, " $F_{Ext.}$ " represents net external force on the system of particles. But, the individual forces on the right hand side represent all forces i.e. both internal and external forces. This means that:

$$F_{Ext.} \quad F_1 + F_2 + \ldots\ldots\ldots + F_n = F_{Ext.} + F_{Int.}$$

It is not difficult to resolve this apparent contradiction. Consider the example of six billiard balls, one of which is moving with certain velocity. The moving ball may collide with another ball. The two balls after collision, in turn, may collide with another ball and so on. The point is that the motions (i.e. velocity and acceleration) of the balls in this illustration are determined by the "internal" contact forces.

Momentum of a system of particles

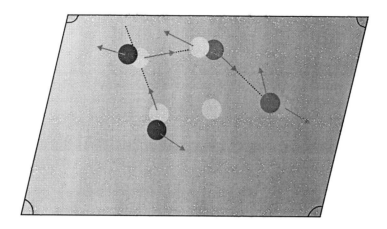

**Figure 6.2: Particles moving with different velocities.**

It happens (law of nature) that the motion of the "center of mass" of the system of particles depend only on the external force - even though the motion of the constituent particles depend on both "internal" and "external" forces. This is the important distinction as to the roles of external and internal forces. Internal force is not responsible for the motion of the center of mass. However, motions of the particles of the system are caused by both internal and external forces.

We again look at the process involved in the example of billiard balls. The forces arising from the collision is always a pair of forces. Actually all force exists in pair. This is the fundamental nature of force. Even the external force like force due to gravity on a projectile is one of the pair of forces. When we study projectile motion, we consider force due to gravity as the external force. We do not consider the force that the projectile applies on Earth. We think projectile as a separate system. In nutshell, we consider a single external force with respect to certain object or system and its motion.

However, the motion within a system is all inclusive i.e both pair forces are considered. It means that internal forces always appear in equal and opposite pair. The net internal force, therefore, is always zero within a system.

$F_{Int.}=0$

This is how the ambiguity in the relation above is resolved.

## Conservation of linear momentum

If no external force is involved, then change in linear momentum of the system is zero:

$F_{Ext.}=0$

$dP=0$

The relation that has resulted is for an infinitesimally small change in linear momentum. By extension, a finite change in the linear momentum of the system is also zero,

$$\Delta P = 0$$

This is what is called "conservation of linear momentum". We can state this conservation principle in following two different ways.

## Definition 2: Conservation of linear momentum

When external force is zero, the change in the linear momentum of a system of particles between any two states (intervals) is zero.

Mathematically,

$$\Delta P = 0$$

We can also state the law as :

## Definition 3: Conservation of linear momentum

When external force on a system of particles is zero, the linear momentum of a system of particles cannot change.

Mathematically,

$$P_i = P_f$$

In expanded form,

$$m_1 v_{1i} + m_2 v_{2i} + \ldots\ldots\ldots\ldots + m_n v_{ni} = m_1 v_{1f} + m_2 v_{2f} + \ldots\ldots\ldots\ldots + m_n v_{nf}$$

We can interpret the conservation of linear momentum in terms of the discussion of billiards balls. The system of billiard balls has certain linear momentum, say, "Pi", as one of the balls is moving. The ball, then, may collide with other balls. This process may repeat as well. The contact forces between the balls may change the velocities of individual balls. However, these changes in velocities take place in magnitude and direction such that the linear momentum of the system of particles remains equal to "Pi".

Momentum of a system of particles

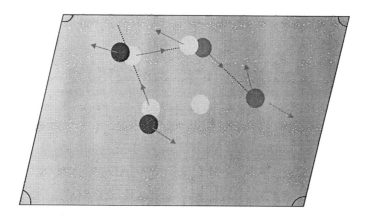

**Figure 6.3: Particles moving with different velocities.**

This situation continues until one of the billiard balls strikes the side of the table. In that case, the force by the side of the table constitutes external to the system of six billiard balls. Also, we must note that when one of the ball goes into the corner hole, the linear momentum of the system changes. This fact adds to the condition for the conservation of linear momentum. Conservation of linear momentum of a system of particles is, thus, subjected to two conditions:

1. No external force is applied on the system of particles.

2. There is no exchange of mass "to" or "from" the system of particles.

In short, it means that the system of particles is a closed system. The definition of the conservation principle is, thus, modified as:

## Definition 4: Conservation of linear momentum

The linear momentum cannot change in a closed system of particles.

# Example 1

*Problem* : A bullet of mass "m" , moving with a velocity "u", hits a wooden block of mass "M", placed on a smooth horizontal surface. The bullet passes the wooden block and emerges out with a velocity "v". Find the velocity of the bullet with respect to the wooden block.

### Bullet piercing through a wooden block

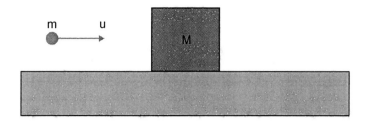

**Figure 6.4: A bullet of mass "m" , moving with a velocity "u", hits a wooden block of mass "M".**

*Solution* : To determine relative velocity of the bullet with respect to wooden block, we need to find the velocity of the wooden block. Let the velocity of block be " $v_1$ ", then relative velocity of the bullet is :

### Bullet piercing through a wooden block

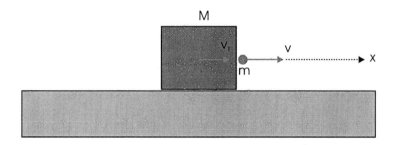

**Figure 6.5: The bullet passes the wooden block and emerges out with a velocity "v".**

$v_{rel} = v - v_1$

The important aspect of the application of conservation of linear momentum is that we should ensure that the system of particles/ bodies is not subjected to net external force (or component of net external force) in the direction of motion.

We note here that there is no external force in horizontal direction on the system comprising of bullet and wooden block. Therefore, we can find the velocity of the wooden block, using conservation of linear momentum.

$\Delta P = mv + mv_1 - mu = 0$

$v_1 = \dfrac{m(u-v)}{M}$

and

$v_{rel} = v - \dfrac{m(u-v)}{M}$

$v_{rel} = \dfrac{Mv}{m} - \dfrac{m(u-v)}{M}$

# Conservation of linear momentum in component form

Linear momentum is a vector quantity. It, then, follows that we can formulate the conservation principle in three linear directions. In three mutually perpendicular directions of a rectangular coordinate system:

$P_{xi} = P_{xf}$

$P_{yi} = P_{yf}$

$P_{zi} = P_{zf}$

These are important results. In most of the situations, we are required to analyze motion in linear direction i.e. one direction. In that case, we are required to analyze force, momentum, etc. in that direction only. This is a great simplification of the analysis paradigm as we shall find out in solving problems.

# Example 2

*Problem* : A ball of mass "m", which is moving with a speed " $v_1$ " in the x-direction, strikes another ball of mass "2m", placed at the origin of the horizontal planar coordinate system. The lighter ball comes to rest after the collision, whereas the heavier ball breaks in two equal parts. One part moves along the y-axis with a speed " $v_2$ ". Find the direction of the motion of the other part.

**Collision of balls**

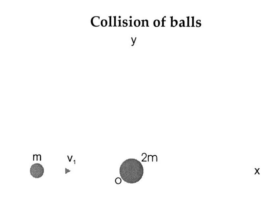

Figure 6.6: A ball collides with another ball at the origin of coordinate system.

*Solution*: The answer to this question makes use of the component form of conservation law of momentum. Linear momentum before the collision is,

$P_{xi} = mv_1$

$P_{yi}=0$

Let the second part of the heavier ball move with a speed "$v_3$" at an angle "$\theta$" as shown in the figure. Linear momentum in two directions after collision,

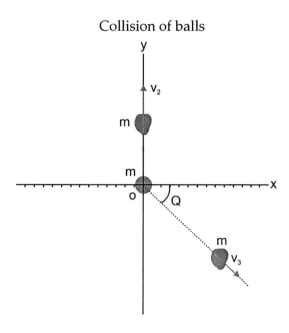

Collision of balls

Figure 6.7: The ball at the origin breaks up in two parts.

$P_{xf}=mv_3\cos\theta$

$P_{yf}=mv_2-mv_3\sin\theta$

Applying the conservation of linear momentum:

$mv_3\sin\theta=mv_1$

and

$mv_2-mv_3\sin\theta=0$

$mv_3\sin\theta=mv_2$

Taking ratio, we have:

$\text{Tan } \theta = \dfrac{v2}{v1}$

$\theta = \tan^{-1}(\dfrac{v2}{v1})$

Collision of objects is a brief phenomenon of interaction with force of relatively high magnitude. It is a common occurrence in the physical world. Atoms, molecules and basic particles frequently collide with each other. An alpha particle, for example, is said to collide when the nucleus of a gold atom repels the incoming alpha particle with a strong electrostatic force. The "contact" is an assessment of the event involving force of high magnitude for a small period. Two objects of macroscopic dimensions are said to collide making physical contact, for example when a tennis ball is hit hard to rebound from a racquet.

**Collision**

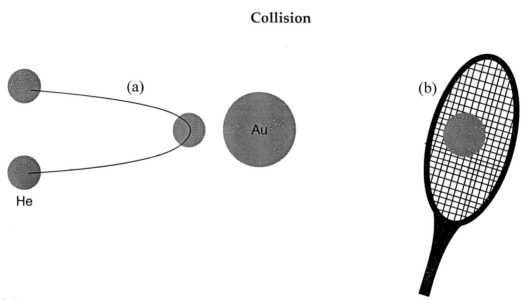

**Figure 6.8**

(a) Alpha particle and nucle us of gold atom come very close to each other. (b) Tennis racquet and ball collide making contact with each other.

Two stationary objects in a given reference cannot collide. Therefore, one of the basic requirements of collision is that at least one of the objects must be moving. Since motion is involved, the system of colliding objects has certain momentum and kinetic energy to begin with. Our objective in this module is to investigate these physical quantities during different states associated with a collision process. A collision is typically studied in terms of three states in "time" reference: (i) before (ii) during and (iii) after.

The state of contact - actual or otherwise - spans a very brief period. The total duration of contact between a tennis racquet and ball may hardly sum up to 30 seconds or so in a set of a tennis match, which might involve hundreds of hits! Typically the contact period during a single collision is milliseconds only. This poses a serious problem as to the measurement of physical quantities during collision. We shall see that the study of collision is basically about making an assessment of happenings in this brief period in terms of quantities that can be measured before and after the collision.

Further, a collision involves forces of relatively high magnitude. To complicate the matter, the force during collision period is a variable force. It makes no sense to attempt investigate the nature of this force quantitatively as it may not be possible to correlate the force with the motions of colliding objects using force law for this very small period. There is yet another reason why we would avoid force analysis during collision. We must understand that a collision need not end up with change in motion only. It is entirely possible that a collision may result in the change in shape (deformation) or even decomposition of the colliding bodies. Such changes resulting from force is not the domain of analysis of motion here.

It is, therefore, imperative that we take a broader approach involving momentum and kinetic energy, to study the behavior of objects when they collide. However, before we proceed to develop the theoretical framework for the collision, it would be helpful to enumerate the characterizing aspects of a collision as:

- A collision involves interaction of an object with force.

- A collision involves forces of relatively high magnitude.

- Collision takes place for a very small period, usually in milliseconds or less.

- A collision involves certain momentum and kinetic energy to begin with.

- The study of collision is not suited for force analysis, using laws of motion.

## Forces during collision

Collision usually occurs in the absence of any external force or comparatively negligible external force with respect to collision force. Two bodies collide because a body comes in the line of motion of another body. The forces during collision between colliding bodies are essentially internal forces to the system of colliding bodies. The internal force is a variable force of relatively high magnitude.

**Force during collision**

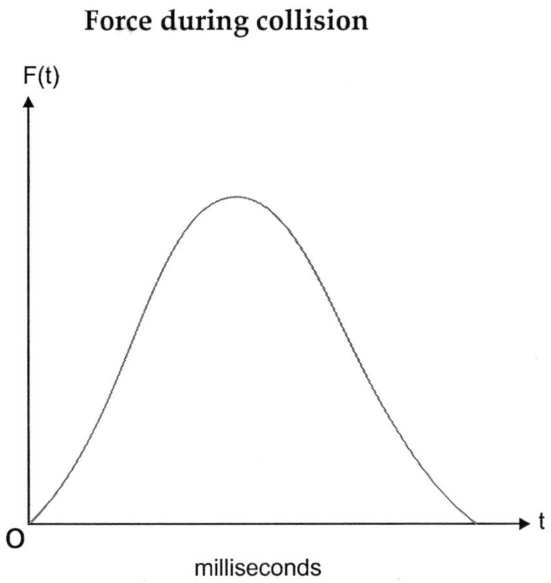

**Figure 6.9: The internal force is a variable force of relatively high magnitude, which operates for a very small period.**

The figure here captures the attributes of the collision force. It is important to note that the plot has time in milliseconds.

The collision forces between colliding objects appear in pairs and follow Newton's third law. The force pairs are equal and opposite. This has important implications. Irrespective of the masses or sizes of the colliding bodies, the pair (two) forces are always equal at any time during the collision. What it means is that the force-time plot of either of the forces is an exact replica of the force-time plot of the other collision force.

Forces during collision

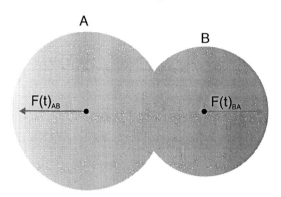

**Figure 6.10: The force pairs are equal and opposite.**

# Impulse

As pointed out earlier, it is not very useful studying contact force, as it is a variable force and secondly, we cannot measure the same with any accuracy for the short interval it operates. However, we can relate the collision force to the change in linear momentum. Here is a word of caution. If the collision forces are internal forces, then how can there be a change in the linear momentum? We are actually referring to change in linear momentum of one of the colliding bodies, which changes in such a manner such that the linear momentum of the system of colliding bodies remains constant. Now, using Newton's second law of motion, the collision force operating on one of the colliding objects is:

$$F(t) = \frac{dp}{dt} \qquad dp = F(t)dt$$

Integrating for the period of collision,

$$\int dp = \int F(t)dt$$

$$\Delta p = \int F(t)dt$$

This is an important result. This enables us to measure the product of two immeasurable quantities (i) variable force of relatively high magnitude, F(t) and (ii) very small time, dt, during the collision in terms of measurable change in momentum, $\Delta p$, before and after the collision. The integral of the product of force and time equals change in momentum and is known as impulse denoted by the symbol "J":

$$J = \Delta p = \int F(t)dt$$

Like linear momentum, impulse is a vector quantity. Evidently, it has the same dimension and unit as that of linear momentum. It may be noted here that impulse is a measure of the product of force and time of operation, but not either of them individually. Further, it must also be understood that impulse is rarely evaluated using its defining integral; rather it is evaluated as the change in linear momentum - most of the time.

We shall learn that a collision may be elastic (where no loss of kinetic energy of the system takes place) or inelastic (where certain loss of kinetic energy of the system is involved).

The collision forces are normal to the tangent through the surfaces in contact. In real situations, the force may not be passing through a contact point, but over an area. In order to keep the analysis simple, however, we consider that contact is a point and the collision forces are normal to the tangent. This simplifying assumption has important implications as we shall see while studying inelastic collision in a separate module.

# Example 1

*Problem* : A ball of 0.1 kg in horizontal flight moves with a speed 50 m/s and hits a bat held vertically. The ball reverses its direction and moves with the same speed as before. Find the impulse that acts on the ball during its contact with the bat.

*Solution* : We calculate the impulse on the ball by measuring the change in linear momentum. Here, vectors involved are one dimensional. Therefore, we can use scalar representation of quantities with appropriate sign. Let us consider that the ball is moving in positive x-direction. Then,

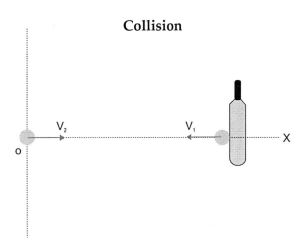

**Figure 6.11: The ball reverses its direction with same speed after its hits the bat.**

$J = \Delta p = p_f - p_i = m v_f - m v_i$

$J = m(v_f - v_i) = 0.1 \times (-50 - 50) = -10$ kgm/s

Negative sign means that the impulse is acting in the "-x" direction, in which the ball finally moves.

# Component form of impulse

Like linear momentum, impulse is a vector quantity. Impulse vector on a body is directed in the direction of $\Delta p$. This direction may not be the direction of either initial or final linear momentum of the body under investigation. As stated earlier in the course, the direction of a vector and that of "change in vector" may not be same. The component of impulse vector along any of the coordinate directions is equal to the difference of the components of linear momentum in that direction. Mathematically,

$$J_x = \Delta p_x = p_{fx} - p_{ix}$$

$$J_y = \Delta p_y = p_{fy} - p_{iy}$$

$$J_z = \Delta p_z = p_{fz} - p_{iz}$$

# Example 2

*Problem* : A ball of 0.1 kg in horizontal flight moves with a speed 50 m/s and is deflected by a bat as shown in the figure at an angle with horizontal. As a result, the ball is reflected in a direction, making 45° with the horizontal. Find the impulse that acts on the ball during its contact with the bat.

**Collision**

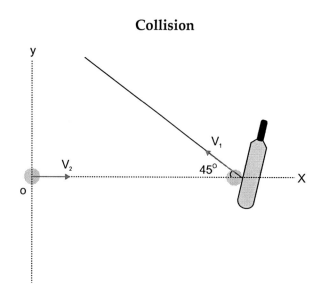

**Figure 6.12: The ball is deflected at an angle with horizontal.**

*Solution* : We calculate the impulse on the ball by measuring the change in linear momentum in the horizontal (x-direction) and vertical (y-direction) directions. Let us consider that the ball is initially moving in the positive x-direction. Then,

$$J_x = \Delta p_x = p_{fx} - p_{ix} = m(v_{fx} - v_{ix})$$

$$\Rightarrow J_x = 0.1 \times (-50\cos45^0 - 50) = -8.54 \text{ kgm/s}$$

Similarly, the component of impulse in the y-direction is:

$$J_y = \Delta p_y = p_{fy} - p_{iy} = m(v_{fy} - v_{iy})$$

$$\Rightarrow J_y = 0.1 \times (50\sin45^0 - 0) = 3.54 \text{ kgm/s}$$

The impulse on the ball is:

$J=J_xi+J_yj$

$J=(-8.54i+3.54j)$ kgm/s

This example and the one given earlier illustrate that the impulse and its evaluation in terms of linear momentum is independent of whether the collision is head-on or not. No doubt the result is different in two cases. The impulse force is greater for a head-on collision than when the colliding object glances the target at an angle. It is expected also. The change in linear momentum, when the striking object reverses its motion, is maximum and as such impulse imparted is also maximum when the collision is head- on.

## Average force

We may not be able to relate impulse with the instantaneous force during the collision, but we can estimate the average force if we have an accurate estimate of the time involved during collision. For the sake of simplicity, we consider one-dimensional contact force. Looking at the expression of impulse i.e. the integral of force over the time of collision is graphically equal to the area under the curve.

**Force - time plot during collision**

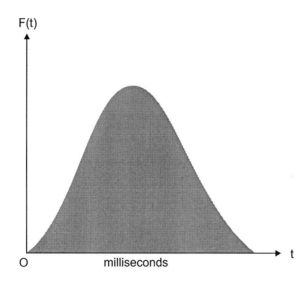

**Figure 6.13: The integral of force over the time of collision is graphically equal to the area under the curve.**

$J=\int F(t)dt=$Area

We can draw a rectangle such that its area is equal to the integral. In such a case the ordinate on the force axis represents the average force during the collision period.

**Average collision force**

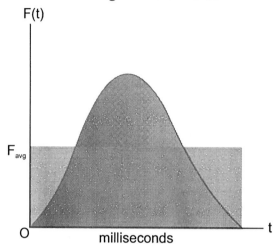

Figure 6.14: Average force is equal to the ordinate of the rectangle of area equal to that of area under the force - time actual curve.

Mathematically,

$$F_{avg} = \frac{\int F(t)dt}{\Delta t} = \frac{J}{\Delta t} = \frac{dP}{dt}$$

# Example 3

*Problem* : A ball of 0.1 kg in horizontal flight moves with a speed 50 m/s and it hits a bat held vertically. The ball reverses its direction and moves with the same speed as before. If the time of contact is 1 millisecond, then find the average force that acts on the ball during its contact with the bat.

*Solution* : We calculate the impulse on the ball by measuring the change in its linear momentum. Let us consider that the ball is moving in the  positive x-direction. Then,

**Collision**

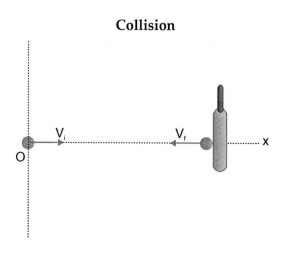

Figure 6.15: The ball reverses its direction with same speed after it hits the bat.

$J=\Delta p=p_f-p_i=mv_f-mv_i$

$J=m(v_f-v_i)=0.1\times(-50-50)=-10$

The magnitude of average force is:

$F_{avg}=10000N$

A collision may take place when external force is operating like when a cricket ball strikes the racket in the presence of gravitational force. However, gravitational force is relatively small (mg = 0.1 x 10 = 1.5 N). This force is negligible in comparison to the collision force of 10000 N! It is, therefore, realistic to assume that collision takes place in the absence of external force.

This concept of average force allows us to estimate the average force on a target, which is bombarded with successive colliding masses like a stream of bullets or balls, as shown in the figure below. Here, the average force is given as:

### Series Collision

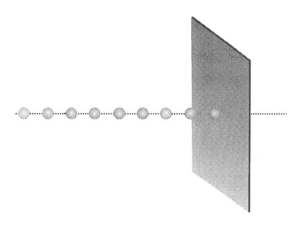

**Figure 6.16: The ball hits the target in quick succession.**

$F_{avg}=\dfrac{J}{\Delta t}=\dfrac{\Sigma\Delta p}{\Delta t}$

If the projectiles have equal mass "m" and there are "n" numbers of the projectiles hitting the target, then

$\Sigma\Delta p=nm\Delta v$

and

$F_{avg}=\dfrac{nm\Delta v}{\Delta t}$

# Linear momentum during collision

We have noted in the examples given here that we can neglect external force for all practical purpose as far as the collision is concerned. Collision force is too large in comparison to forces like gravitational force. For this reason, we shall neglect external forces unless stated otherwise. As such, linear momentum of the system of colliding objects is conserved. This means that the linear momentum of the system before and after the collision is same.

$P_{before} = P_{after}$

This fact provides the basic framework for analyzing collision. It is valid irrespective of the type of collision: elastic or inelastic. We shall make use of this fact in subsequent modules, while analyzing various scenarios of collision.

# Kinetic energy of colliding bodies

There are two types of collision - elastic and inelastic. In the elastic collision, the colliding bodies are restored to the normal shape and size when the bodies separate. The deformation of colliding bodies is temporary and is limited to the period of collision. This means that kinetic energy of the system is simply stored as elastic energy during collision and is released to the system when bodies regain their normal condition. On the other hand, there is loss of kinetic energy during inelastic collision as kinetic energy is converted to some other form of energy like sound energy or heat energy. Such energy cannot be released to the system of colliding bodies as kinetic energy subsequent to the collision.

In reality most of the macroscopic collisions involving physical contact are inelastic, as kinetic energy of the system is irrevocably converted to other forms of energy. For example, when a tennis ball is released from a height, it does not bounce to the same height on the return journey. However, if the ground is plane and hard, then the ball may transverse the greater part of the height and we may say that the collision is approximately elastic.

There is an interesting set up, which explains various aspects of elastic collision. We will study this set up, which comprises of two blocks and a spring to get an insight into the collision process. We imagine two identical blocks "A" and "B" moving on a smooth floor as shown in the figure below. The block "A" moves right with speed $v_A$ towards block "B". On the other hand block "B" moves with speed $v_B$ such that $v_A > v_B$. We also imagine that a "mass-less" spring is attached to the block "B" as shown in the figure. Let us also consider that the spring follows Hook's law for elongation (F = -kx).

## Collision

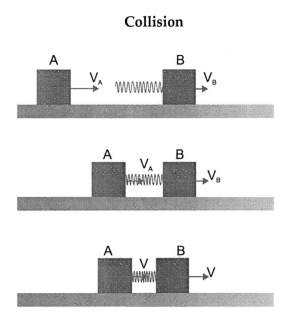

**Figure 6.17: The spring is compressed.**

At the time when block "A" hits the spring, the end of the spring in contact with block "A" acquires the speed of the block "A" ( $v_A$ ). The end of the spring in contact with block "B", however, moves with speed $v_B$. As the block "A" moves with relative speed towards "B", the spring is compressed. The spring force for any intermediate compression "x" is given by:

$F(t)=-kx$

The spring forces act on both the blocks but in opposite directions. The spring force is equivalent to the internal collision force. The spring force pushes the block "A" towards the left and hence decelerates it. Let its speed at a given instant after hitting the spring be $v_A'$. On the other hand, spring force pushes the block "B" towards the right and hence accelerates it. As such, block "B" acquires certain speed, say $v_B'$. If $v_A' > v_B'$, then the spring is further compressed. The speed of block "A" further decreases, and that of block "B" further increases. The spring force during this period keeps increasing. This process continues until the speeds of blocks become equal. Let us consider that common speed of each of the blocks be V.This situation corresponds to the maximum compression and maximum spring force i.e the maximum collision force.

The maximum spring force continues to decelerate block "A" and accelerates block "B". As such, block "B" ( $v_B''$ ) begins to move faster than the block "A" ( $v_A''$ ) . It results in elongation of spring. The spring force during this period keeps decreasing. The spring force, however, continues to decelerates block "A" and accelerates block "B". This process continues until the spring attains its normal length. At this moment, the block "A" looses contact with the spring. The block "B" moves with greater speed ( $v_B'''$ ) than block "A" ( $v_A'''$ ). Therefore, the separation between the two blocks keeps increasing with time as they move with different speeds.

**Collision**

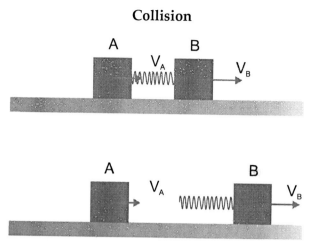

**Figure 6.18: The spring is elongated.**

The spring force - time plot during the contact with spring approximates the force curve during a collision:

**Collision force**

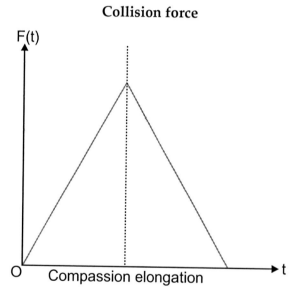

**Figure 6.19: The force - time plot.**

Since the spring is mass-less and elastic (follows Hooke's law), the kinetic energy of the system before the collision is temporarily converted to elastic potential energy i.e spring energy during the compression of the spring. The stored elastic potential energy is then released as kinetic energy during elongation of the spring. At the end of elongation when spring attains its normal length, the kinetic energy of the system is restored as before. In a nutshell,

$$K_{\text{before}} = K_{\text{after}}$$

During collision, however, total energy has a component of elastic potential energy as well, and is equal to the kinetic energy of the system before and after the collision:

$E=K+U=K_{before}=K_{after}$

In the same fashion, we can visualize inelastic collision. In this case, some of the energy is converted to a form of energy which cannot be regained as kinetic energy, like when a rubber ball hits a ground and is unable to regain the height. Here,

$K_{before}>K_{after}$

# Summary

1: Collision of objects is a brief phenomenon of interaction with force of relatively high magnitude.

2: The integral of the product of force and time equals change in momentum and is known as impulse denoted by the symbol "$J$" :

$J=\Delta p=\int F(t)dt$

3: The component of impulse vector along any of the coordinate directions is equal to the difference of the components of linear momentum in that direction. Mathematically,

$J_x=\Delta p_x=p_{fx}-p_{ix}$

$J_y=\Delta p_y=p_{fy}-p_{iy}$

$J_z=\Delta p_z=p_{fz}-p_{iz}$

4: The average force during collision is by the ratio of impulse and time interval as:

$F_{avg}=\dfrac{\int F(t)dt}{\Delta t}\dfrac{J}{\Delta t}=\dfrac{\Delta P}{\Delta t}$

5: Elastics collision:

$P_{before}=P_{after}$

and

$K_{before}=K_{after}$

6: Inelastic collision:

$P_{before}=P_{after}$

and

$K_{before}>K_{after}$

For applications of momentum and collisions, please visit:

http://cnx.org/content/m14854/latest/

# PowerPoint Link:

Please refer to the end of the module link.

# Discussion Question

Answer the following questions with references. Please remember to follow the standard APA referencing style.

For APA standards of references, please visit: http://owl.english.purdue.edu/owl/resource/560/01/

Also, respond in detail to one other post by fellow students.

**6.1** One can stop a mosquito travelling at a certain speed towards you by intercepting it with your hand. Can you do the same for a car moving at the same speed as a mosquito? Discuss this scenario using your knowledge on momentum and impulse. Give examples of calculations to elaborate your answer.

**6.2** Explain the energy loss in a glancing collision with suitable examples and values. Compare this against head-on collisions.

# Laboratory Activity and the link

Go to the link below and run the simulation:

http://phet.colorado.edu/simulations/sims.php?sim=Forces_in_1_Dimension

Go to: http://phet.colorado.edu/en/contributions/view/2891


Open onedimensional forces.doc

Do the activity.

# Module 4 Lecture Links:

http://ocw.mit.edu/courses/physics/8-01-physics-i-classical-mechanics-fall-1999/video-lectures/lecture-15/

http://ocw.mit.edu/courses/physics/8-01-physics-i-classical-mechanics-fall-1999/video-lectures/lecture-16/

http://ocw.mit.edu/courses/physics/8-01-physics-i-classical-mechanics-fall-1999/video-lectures/lecture-17/

# Module 4 – Student's self-assessment

**Please answer the following questions, which give you an indication of your standard of learning for this module.**

- Could you describe momentum?

- Could you describe impulse?

- Can you apply knowledge to solve problems involving momentum and impulse?

- What sections of the textbook would you read for more information on the above?

- Did you enjoy your lesson?

- What aspect of the lesson was most interesting to you?

- Can you now confidently do the lesson's activities?

**Now take the chapter 6 test (not included with this book)**

# Module 5

## CHAPTER-7

# Rotational Motion and the Law of Gravity

## Objectives

At the end of this lesson, you should be able to:

1.  Answer questions on circular motions
2.  Apply angular speed, velocity, acceleration equations to solve problems

## Lecture Notes

### Uniform Circular Motion

Consider an object moving along a circular path with constant or uniform speed.

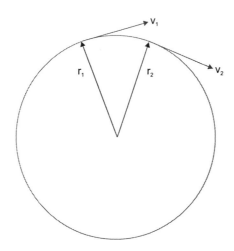

Figure 7.1

When an object travels in a circle, the **direction** of its velocity changes. The **speed remains uniform,** the **velocity changes** because its direction changes.

$$\Delta \mathbf{v} = \mathbf{v}_2 - \mathbf{v}_1$$

This is **not zero**, yet:

$$\mathbf{v}_2 \mid = v_2 = \mid \mathbf{v}_1 \mid = v_1 = v$$

v is the **speed**, which remains constant. Observe $\Delta \mathbf{v}$ from the following diagram:

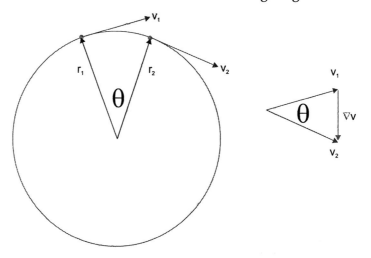

**Figure 7.2**

Notice the two **similar triangles** in the position vectors and the velocity vectors,

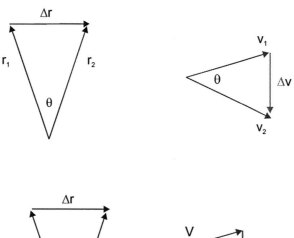

or

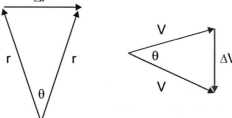

or

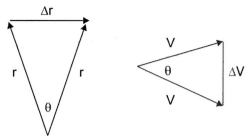

Since these are similar triangles,

$\Delta v / v = \Delta r / r$

For small changes in time or for small angles $\theta$, the distance $\Delta r$ is very nearly given by

$\Delta r = v \, \Delta t$

This means

$\Delta v / v = \Delta r / r$

$\Delta v / v = v \, \Delta t / r$

$\Delta v = v^2 \Delta t / r$

$\Delta v / \Delta t = v^2 / r$

$\Delta v / \Delta t$ is the acceleration an object has because it moves in a circle; this is called the **centripetal acceleration** $a_c$ and is directed toward the center of the circle. Therefore,

$$a_c = \frac{v^2}{r}$$

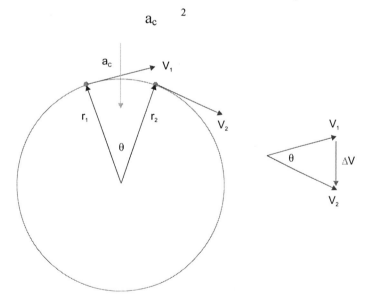

**Figure 7.3**

# Angular velocity

The speed at which something turns, rotates or revolves is **angular velocity** and the symbol is $\omega$. It may have units of revolutions per minute, revolutions per second, or radians per second.

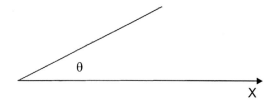 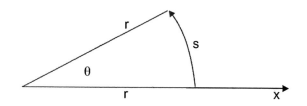

We can measure the angle θ in degrees using a protractor. The other way is to use the following formula. Any angle can be defined as the ratio of the **arc length s** to the **radius r**; that is,

$$\theta = \frac{S}{r}$$

We call this unit of angular measure a **radian**.

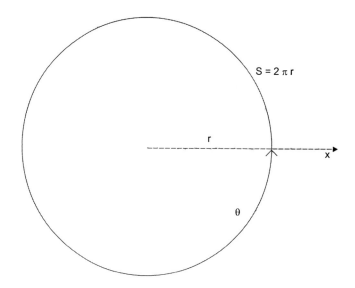

Consider a complete circle. We would describe a complete circle as having an angle of 360°. In terms of radians, a complete circle has an arc length equal to its circumference, s = C = 2πr.

$$\theta = \frac{2\pi r}{r} = 2\pi$$

$$\theta = \frac{2\pi r}{r} = 2\pi \text{ radians}$$

$$2\pi \text{ radians} = 360^0$$

$$1 \text{ radians} = 57.3^0$$

The arc length traveled by a point on a rotating object is equal to the radius of that point multiplied by the angle θ through which it has rotated,

$$s = r\,\theta$$

where θ is measured in radians,

That means the linear speed v of a point on a rotating object is equal to the radius of that point multiplied by the angular speed ω with which it is rotating, provided ω is measured in radians per time. That is,

v = rω

provided ω is measured in **radians** per **time**.

- The **net force** on a car traveling around a curve is the **centripetal force**, $F_c = m\ v^2\ /\ r$, directed toward the center of the curve.

- For a **level curve**, the centripetal force will be supplied by the friction force between the tires and roadway.

- A **banked curve** can supply the centripetal force by the normal force and the weight without relying on friction.

We have already studied the kinematics of circular motion. Specifically, we observed that a force, known as centripetal force, is required for a particle to execute circular motion.

Force is required because the particle executing circular motion needs to change direction continuously. In the case of uniform circular motion (UCM), speed of the particle is constant. The change in velocity is only in terms of change in direction. This needs a force in the radial direction to meet the requirement of the change in direction continuously.

The acceleration is given by:

$$a_r = \frac{v^2}{r} = \omega^2 r$$

The centripetal force is given by:

$$F_C = ma_r = \frac{mv^2}{r} = m\omega^2 r$$

**Centripetal force**

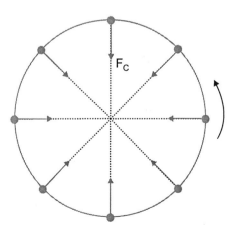

**Figure 7.4: Uniform circular motion requires a radial force that continuously change its direction, while keeping magnitude constant.**

## Centripetal force

The specific requirement of a continuously changing radial force is not easy to meet by mechanical arrangement. The requirement means that force should continuously change its direction along with particle. It is a tall order. Particularly, if we think of managing force by physically changing the mechanism that applies force. Fortunately, natural and many craftily thought out arrangements create situations in which the force on the body changes direction with the change in the position of the particle - by the very act of motion. One such arrangement is the solar system, in which gravitational force on the planet is always radial.

Centripetal force is a name given to the force required for circular motion. The net component of external forces which meet this requirement is called centripetal force. In this sense, centripetal force is not a separately existing force. Rather, we should look at this force as a component of the external forces on the body in the radial direction.

## Direction of centripetal force and circular trajectory

There is a subtle point about circular motion with regard to the direction of force as applied on the particle in circular motion. If we apply force on a particle at rest, then it moves in the direction of applied force and not perpendicular to it. In circular motion, the situation is different. We apply force (centripetal) to a particle, which is already moving in a direction perpendicular to the force. As such, the resulting motion from the interaction of motion with external force is not in the radial direction, but in a tangential direction.

In accordance with Newton's second law of motion, the particle accelerates along the direction of centripetal force i.e. towards the center. As such, the particle actually transverses a downward displacement ($\Delta y$) with centripetal acceleration; but at the same time, the particle moves sideways ($\Delta x$) with constant speed, as the component of centripetal force in the perpendicular direction is zero.

It may sound odd, but the fact is that the particle is continuously falling towards the center in the direction of centripetal force and at the same is able to maintain its linear distance from the center, owing to constant side way motion.

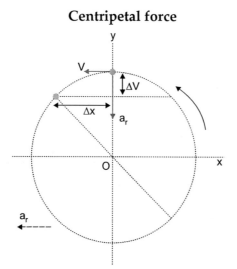

**Figure 7.5: Direction of centripetal force and circular trajectory**

In the figure shown, the particle moves towards center by $\Delta y$, but in the same period the particle moves left by $\Delta x$. In the given period, the vertical and horizontal displacements are such that resultant displacement finds the particle always on the circle.

$$\Delta x = v\Delta t$$

$$\Delta y = \frac{1}{2}ar\Delta t2$$

## Force analysis of uniform circular motion

As pointed our earlier, we come across large numbers of motion, where natural setting enables continuous change of force direction with the moving particle. We find that a force meeting the requirement of centripetal force can be any force type like friction force, gravitational force, tension in the string or electromagnetic force. Here, we consider some of the important examples of uniform circular motion drawn from our life experience.

## Uniform circular motion in horizontal plane

A particle tied to a string is rotated in a horizontal plane by virtue of the tension in the string. The tension in the string provides the centripetal force for uniform circular motion.

However, we should understand that this force description is actually an approximation, because it does not take into account the downward force due to gravity. In fact, it is not possible to have a horizontal uniform circular motion (except in the region of zero gravity) by keeping the string in horizontal plane. It is so because; gravitational pull will change the plane of string and the tension in it.

In order that there is horizontal uniform circular motion, the string should be slanted such that the tension as applied to the particle forms an angle with the horizontal plane. The horizontal component of the tension provides the needed centripetal force, whereas the vertical component balances the weight of the particle.

**Uniform circular motion in horizontal plane**

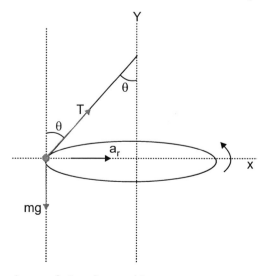

**Figure 7.6: String is not in the plane of circular motion.**

$$\sum F_x \Rightarrow T\sin\theta = ma_r = \frac{mv^2}{r}$$

and

$$\sum F_y \Rightarrow T\cos\theta = mg$$

Taking the ratio,

$$\Rightarrow \tan\theta = \frac{mv^2}{rg}$$

# Example 1

*Problem* : A small boy sits on a horizontal platform of a joy wheel at a linear distance of 10 m from the center. When the wheel exceeds 1 rad/s, the boy starts slipping. Find the coefficient of friction between the boy and the platform.

*Solution* : For the boy to be stationary with respect to the platform, forces in both vertical and horizontal directions are equal. However, the requirement of centripetal force increases with increasing rotational speed. If centripetal force exceeds the maximum static friction, then the boy begins to slip towards the center of the rotating platform.

**Horizontal circular motion**

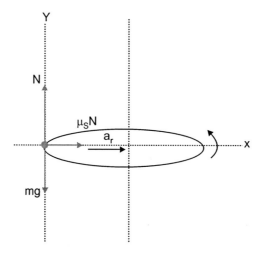

**Figure 7.7**

In the vertical direction,

$$N=mg$$

In the horizontal direction,

$$\Rightarrow m\omega^2 r = \mu_s N = \mu_s mg$$

$$\Rightarrow \mu_s = \frac{r\omega 2}{g} = 5 \, X \frac{1^2}{10} = 0.5$$

# Motion of a space shuttle

A space shuttle moves in a circular path around Earth. The gravitational force between Earth and the shuttle provides for the centripetal force:

$$mg = \frac{mv^2}{r}$$

Where "g" is the acceleration due to gravity (acceleration arising from the gravitational pull of Earth) on the satellite.

Here, we need to point out an interesting aspect of centripetal force. A person is subjected to centripetal force while moving in a car and as well when moving in a space shuttle. However, the experience of the person in the two cases is different. In the car, the person experiences (feels) a normal force in the radial direction as applied to a part of the body. On the other hand, a person in the shuttle experiences the "feeling" of weightlessness. Why this difference when the body experiences centripetal force in either case?

In the space shuttle, gravity acts on each of the atoms constituting our body and this gravity itself is the provider of centripetal force. There is no push on the body as in the case of car. The body experiences the "feeling" of weightlessness as both space shuttle and the person are continuously falling towards the center of Earth. The person is not able to push other bodies. Importantly, gravitational pull or weight of the person is equal to mg and not equal to zero.

# Horizontal circular motion in a rotor

Horizontal rotor holds an object against the wall of a rotating cylinder at a certain angular speed. The object (which could be a person in a fun game arrangement) is held by friction between the surfaces of the object and the cylinder's inside wall. For a given weight of the object, there is a threshold minimum velocity of the rotor (cylinder); otherwise the object will fall down.

The object has a tendency to move straight. As the object is forced to move in a circle, it tends to move away from the center. This means that the object presses the wall of the rotor. The rotor, in turn, applies normal force on the object towards the center of circular path.

$$\sum F_x = N = ma_r = \frac{mv^2}{r}$$

Since friction is linearly related to normal force for a given pair of surfaces ($\mu_s$), it is possible to adjust the speed of the rotor such that maximum friction is equal to the weight of the object. In the vertical direction, we have:

Horizontal circular motion in a rotor

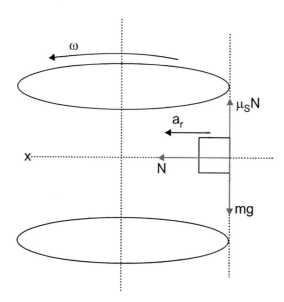

**Figure 7.8: As a limiting case, the maximum friction is equal to the weight of the object.**

$$\sum F_y = \mu_s N - mg = 0$$

$$\Rightarrow N = \frac{mg}{\mu_s}$$

By combining the two equations, we have,

$$\Rightarrow \frac{mv^2}{r} = mg/\mu_s$$

$$\Rightarrow v = \sqrt{(\frac{rg}{\mu_s})}$$

This is the threshold value of speed for the person to remain stuck with the rotor.

We note the following points about the horizontal rotor :

1.    The object tends to move away from the center owing to its tendency to move straight.

2.    A normal force acts towards the center, providing centripetal force

3.    Normal force contributes to maximum friction as $F_s = \mu_s N$.

4.    Velocity of the rotor is independent of the mass of the object.

# Force analysis of non-uniform circular motion

Motion in a vertical loop involves non-uniform circular motion. To illustrate the force analysis, we consider the motion of a cyclist, who makes circular rounds in vertical plane within a cylindrical surface by maintaining a certain speed.

# Vertical circular motion

In the vertical loop within a hollow cylindrical surface, the cyclist tends to move straight in accordance with its natural tendency. The curvature of the cylinder, however, forces the cyclist to move along a circular path (by changing direction). As such, the body has the tendency to press the surface of the cylindrical surface. In turn, the cylindrical surface presses the body towards the center of the circular path.

**Vertical circular motion**

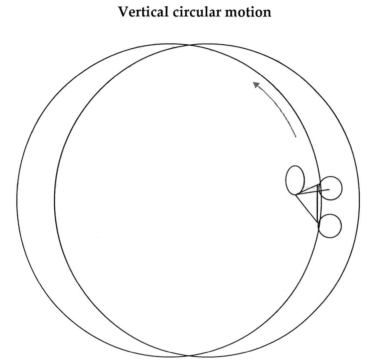

**Figure 7.9: The cyclist executes vertical circular motion along the cylindrical surface.**

The free body diagram of the cyclist at an angle "θ" is shown in the figure. We see that the resultant of normal force and the component of weight in the radial direction meets the requirement of centripetal force in the radial direction,

## Vertical circular motion

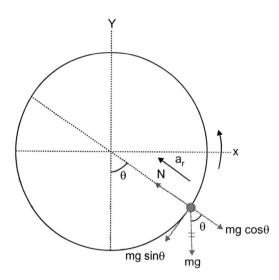

**Figure 7.10 : Force diagram**

$$N-mg\cos\theta = \frac{mv^2}{r}$$

The distinguishing aspects of circular motion in a vertical plane are listed here:

1. Motion in a vertical loop is a circular motion – not uniform circular motion. It is so because there is both radial force (N – mg cosθ) and tangential force (mg sin θ). Radial force meets the requirement of centripetal force, whereas tangential force accelerates the particle in the tangential direction. As a result, the speed of the cyclist decreases while traveling up and increases while traveling down.

2. Centripetal force is not constant, but changing in magnitude as the speed of the cyclist is changing, and is dependent on the angle "θ".

The cyclist is required to maintain a minimum speed to avoid free fall. The possibility of free fall is most stringent at the highest point of the loop. We, therefore, analyze the motion at the highest point with the help of the free body diagram as shown in the figure.

## Vertical circular motion

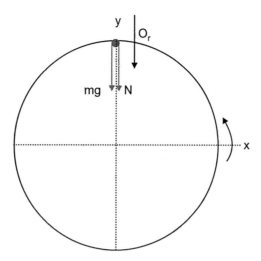

**Figure 7.11: Force diagram at the top**

$$N+mg = \frac{mv^2}{r}$$

*Note*:

We can also achieve the result as above by putting the value θ=180° in the equation obtained earlier.

The minimum speed of the cyclist corresponds to the situation when normal force is zero. For this condition,

$$mg= \frac{mv^2}{r}$$

$$v=\sqrt{(rg)}$$

## Vertical motion of a particle attached to a string

This motion is the same as discussed above. The only difference is that the tension of the string replaces normal force in this case. The force at the highest point is given as:

$$T+mg = \frac{mv^2}{r}$$

Also, the minimum speed for the string not to slack at the highest point (T = 0),

$$mg = \frac{mv^2}{r}$$

$$v=\sqrt{(rg)}$$

The complete analysis of circular motion in a vertical plane involves considering forces on the body at different positions. However, external forces depend on the position of the body in the circular trajectory. The forces are not constant forces as in the case of circular motion in a horizontal plane.

We shall learn subsequently that a situation involving variable force is best analyzed in terms of energy concept. As such, we will revisit vertical circular motion again after studying different forms of mechanical energy.

For detailed applications on several scenarios on circular motions, please visit:

http://cnx.org/content/m13871/latest/

# Gravitational Force

Gravity is an inherent property of all matter. Two bodies attract each other by virtue of their mass. This force between two bodies of any size (an atom or a galaxy) signifies existence of matter and is known as gravitational force.

Gravitational force is the weakest of four fundamental forces. Therefore, it is experienced only when at least one of two bodies has considerable mass. This presents difficulties in setting up illustrations with terrestrial objects. On the other hand, gravitation is the force that sets up our universe and governs motions of all celestial bodies. Orbital motions of satellites – both natural and artificial – are governed by gravitational force.

Newton derived a law to quantify gravitational force between two "particles". The famous incidence of an apple falling from a tree stimulated Newton's mind to analyze observations and carry out series of calculations that finally led him to propose a universal law of gravitation. A possible sequence of reasoning, leading to the postulation is given here:

1:   The same force of attraction works between "Earth (E) and an apple (A)", and between "Earth (E) and Moon (M)".

2:   From the analysis of data available at that time, he observed that the ratio of forces of attraction for the above two pairs is equal to the ratio of square of distance involved as:

**Gravitational force**

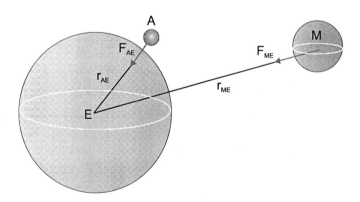

**Figure 7.12: Earth - moon - apple system**

$$\frac{\text{Fme}}{\text{Fae}} = \frac{rae^2}{rme^2}$$

3:    From the above relation, Newton concluded that the force of attraction between any pair of two bodies is inversely proportional to the square of linear distance between them.

$$F \propto \frac{1}{r^2}$$

4:    From the second law of motion, force is proportional to the mass of the body being subjected to gravitational force. From the third law of motion, forces exist in an equal and opposite pair. Hence, gravitational force is also proportional to the mass of another body. Newton concluded that the force of gravitation is proportional to the product of the mass of the two bodies.

$F \propto m_1 m_2$

5:    Combining two "proportional" equations and introducing a constant of proportionality, "G", Newton proposed the gravitational law as:

$$\Rightarrow F = \frac{Gm_1m_2}{r^2}$$

In order to emphasize the universal character of gravitational force, the constant "G" is known as the "Universal gravitational constant". Its value is,

$G = 6.67 \times 10^{-11} N-m^2/kg^2$

This law was formulated based on the observations of real bodies - small (apple) and big (Earth and moon). However, Newton's law of gravitation is stated strictly for two particles. The force pair acts between two particles, along the line joining their positions as shown in the figure:

**Gravitational force**

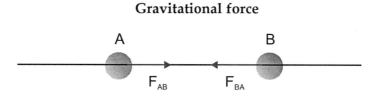

**Figure 7.13: Force between two particles placed at a distance**

Extension of this law to real bodies like the pairing of Earth and Moon can be understood, as bodies are separated by large distances (about $0.4 \times 10^6$ km), compared to the dimension of the bodies (in thousands km). Two bodies, therefore, can be treated as particles.

On the other hand, Earth cannot be treated as a particle for the pairing of "Earth and Apple", based on the reasoning of large distance. The apple is right on the surface of Earth. Newton proved a theorem that every spherical shell (hollow sphere) behaves like a particle for a particle external to it. Extending this theorem, Earth, being a continuous composition of infinite numbers of shells of different radii, behaves - as a whole - like a particle for an external object like an apple.

In fact, we will prove this theorem, employing the gravitational field concept for a spherical mass like that of Earth. For the time being, we consider Earth and the apple as particles, based on the Newton's shell theory. In that case, the distance between Apple and the center of Earth is equal to the radius of Earth i.e. 6400 km.

# Magnitude of force

The magnitude of gravitational force between terrestrial objects is too small to experience. A general question that arises in the mind of a beginner is "why do not we experience this force between, say, a book and pencil?" The underlying fact is that gravitational force is indeed a very small force for masses that we deal with in our immediate surrounding - except Earth itself.

We can appreciate this fact by calculating the force of gravitation between two particle masses of 1 kg each, which are 1 m apart:

$$\Rightarrow F = \frac{6.67 X 10^{-11} X 1 X 1}{1^2}$$

$$F = 6.67 X 10^{-11} N$$

This is too insignificant a force to manifest against bigger forces like the force of gravitation due to Earth, friction, or force due to atmospheric pressure, wind, etc.

This typifies the small value of "G", which renders the force of gravitation so small for terrestrial objects. Gravitation plays a visible and significant role where masses are significant, like that of planets including our Earth, stars and such other massive aggregation, including "black holes" with extraordinary gravitational force to hold back even light. This is the reason, we experience the gravitational force of Earth, but we do not experience the gravitational force due to a building or any such structures on Earth.

# Gravitational force vector

Newton's law of gravitation expresses gravitational force between two bodies. Here, gravitational force is a vector. However, the force vector is expressed in terms of quantities, which are not vectors. The linear distance between two masses, appearing in the denominator of the expression, can have either of two directions from one to another point mass.

Even if we refer to the linear distance between two particles via a reference direction, the vector appears in the denominator and is, then, squared also. In order to express gravitational force in vector form, therefore, we shall consider a unit vector in the reference direction and use the same to denote the direction of force as:

## Direction of gravitational force

**Figure 7.14: Force between two particles placed at a distance**

$$F_{12} = \frac{G m_1 m_2 r}{r^2}$$

$$F_{21} = -\frac{G m_1 m_2 r}{r^2}$$

Note that we need to put a negative sign before the second expression to make the direction consistent with the direction of the gravitational force of attraction. We can easily infer that the sign in the expression actually depends on the choice of threference direction.

## Net gravitational force

Gravitational force is a vector quantity. The net force of gravitation on a particle is equal to the resultant of forces due to all other particles. This is also known as the "superposition principle", in which the net effect is the sum of individual effects. Mathematically,

$$\Rightarrow F = \Sigma F_i$$

Here, $F$ is the net force due to other particles 1, 2, 3, and so on.

An extended body is considered to be a continuous aggregation of elements, which can be treated as particles. This fact can be represented by an integral of all elemental forces due to all such elements of a body, which are treated as particles. The force on a particle due to an extended body, therefore, can be computed as:

**Net gravitational force**

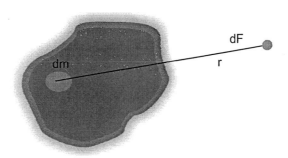

**Figure 7.15: Net gravitational force is vector sum of individual gravitations due to particle like masses.**

$$F = \int dF$$

where integration is evaluated to include all mass of a body.

## Examples

*Problem 1:* Three identical spheres of mass "M" and radius "R" are assembled to be in contact with each other. Find the gravitational force on any of the spheres due to the remaining two spheres. Consider no external gravitational force exists.

Three identical sphered in contact

**Figure 7.16: Three identical spheres of mass "M" and radius "R" are assembled in contact with each other.**

*Solution :* The gravitational forces due to pairs of any two spheres are equal in magnitude, making an angle of 60° with each other. The resultant force is,

Three identical sphered in contact

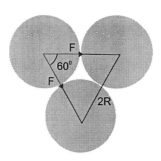

**Figure 7.17: Each sphere is attracted by other two spheres.**

$$\Rightarrow R = \sqrt{(F^2 + F^2 + 2F^2 \cos 60^0)}$$

$$R = \sqrt{3}\,F$$

Now, the distance between centers of mass of any pair of spheres is "2R". The gravitational force is,

$$F = \frac{GM^2}{(2R)^2}$$

Therefore, the resultant force on a sphere is,

$$F = \sqrt{3}\,\frac{GM^2}{4R^2}$$

# Measurement of universal gravitational constant

The universal gravitational constant was first measured by Cavendish. The measurement was an important achievement in the sense that it could measure the small value of "G" quite accurately.

The arrangement consists of two identical small spheres, each of mass "m". They are attached to a light rod in the form of a dumb-bell. The rod is suspended by a quartz wire of known coefficient of torsion "k" such that the rod lies in a horizontal plane. A mirror is attached to the quartz wire, which reflects a light beam falling on it. The reflected light beam is read on a scale. The beam, mirror and scale are all arranged in one plane.

**Cavendish experiment**

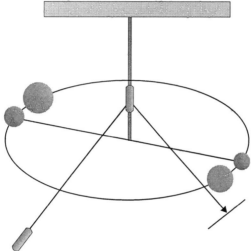

**Figure 7.18: Measurement of universal gravitational constant**

The rod is first made to suspend freely and stabilize at an equilibrium position. As no net force acts in the horizontal direction, the rod should rest in a position without any torsion in the quartz string. The position of the reflected light on the scale is noted. This reading corresponds to the neutral position, when no horizontal force acts on the rod. The component of Earth's gravitation is vertical. Its horizontal component is zero. Therefore, it is important to keep the plane of rotation horizontal to eliminate the effect of Earth's gravitation.

Two large and heavier spheres are then brought across, close to the smaller sphere such that the centers of all spheres lie on a circle as shown in the figure above. The gravitational forces due to each pair of small and big mass, are perpendicular to the rod and opposite in direction. Two equal and opposite forces constitutes a couple, which is given by,

$$\tau_G = F_G L$$

where "L" is the length of the rod.

The couple caused by gravitational force is balanced by the torsion in the quartz string. The torque is proportional to angle "θ" through which the rod rotates about the vertical axis.

$$\tau_T = k\theta$$

The position of the reflected light is noted on the scale for the equilibrium. In this condition of equilibrium,

$$\Rightarrow F_G L = k\theta$$

Now, the expression of Newton's law of gravitation for the gravitational force is:

$$F_G = \frac{GMm}{r^2}$$

Where "m" and "M" are the mass of small and big spheres. Putting this in the equilibrium equation, we have:

$$F = \frac{GMmL}{r^2} = k\theta$$

Solving for "G", we have,

$$\Rightarrow G = r^{2k\theta}/MmL$$

In order to improve the accuracy of the measurement, the bigger spheres are then, placed on the opposite sides of the smaller spheres with respect to earlier positions (as shown in the figure below). Again, the position of reflected light is noted on the scale for the equilibrium position, which should lie opposite to the earlier reading, about the reading corresponding to the neutral position.

# Gravitational Potential Energy

The description of force having "action at a distance" is best described in terms of force field. The "per unit" measurement is a central idea of a force field. The field strength of a gravitational field is the measure of gravitational force experienced by unit mass. On a similar footing, we can associate energy with the force field. We shall define a quantity of energy that is associated with the position of unit mass in the gravitational field. This quantity is called gravitational potential (V) and is different from potential energy as we have studied earlier. Gravitational potential energy (U) is the potential energy associated with any mass - as against unit mass in the gravitational field.

Two quantities (potential and potential energy), though different, are closely related. From the perspective of force field, the gravitational potential energy (U) is the energy associated with the position of a given mass in the gravitational field. Clearly, two quantities are related to each other by the equation,

*U=mV*

The unit of gravitational potential is Joule/kg.

There is a striking parallel among various techniques that we have so far used to study force and motion. One of the techniques employs vector analysis, whereas the other technique employs scalar analysis. In general, we study motion in terms of force (vector context), using Newton's laws of motion, or in terms of energy, employing the "work-kinetic energy" theorem or conservation law (scalar context).

In the study of conservative force like gravitation also, we can study gravitational interactions in terms of either force (Newton's law of gravitation) or energy (gravitational potential energy). It follows, then, that the study of conservative force in terms of "force field" should also have two perspectives, namely that of force and energy. Field strength presents the perspective of force (vector character of the field), whereas gravitational potential presents the perspective of energy (scalar character of field).

# Gravitational potential

The definition of gravitational potential energy is extended to unit mass to define gravitational potential.

## *Definition 1: Gravitational potential*

The gravitational potential at a point is equal to "negative" of the work by the gravitational force as a particle of unit mass is brought from infinity to its position in the gravitational field.

Or

## *Definition 2: Gravitational potential*

The gravitational potential at a point is equal to the work by the external force as a particle of unit mass is brought from infinity to its position in the gravitational field.

Mathematically,

$$V = -W_G = -\int_{\infty}^{r} \frac{fg\,dr}{m} = -\int_{\infty}^{r} E\,dr$$

Here, we can consider the gravitational field strength, "E", in place of gravitational force, " $F_G$ ", to account for the fact we are calculating work per unit mass.

# Change in gravitational potential in a field due to point mass

The change in gravitational potential energy is equal to the negative of work by gravitational force as a particle is brought from one point to another in a gravitational field. Mathematically,

$$\Delta U = -\int_{r1}^{r2} Fg\,dr$$

Clearly, change in gravitational potential is equal to the negative of work by gravitational force as a particle of unit mass is brought from one point to another in a gravitational field. Mathematically:

$$\Rightarrow \Delta V = \frac{\Delta U}{m} = -\int_{r1}^{r2} E\,dr$$

We can easily determine the change in potential as a particle is moved from one point to another in a gravitational field. In order to find the change in potential difference in a gravitational field due to a point mass, we consider a point mass "M", situated at the origin of reference. Considering motion in the reference direction of "r", the change in potential between two points at a distance "r" and "r+dr" is:

## Gravitational potential

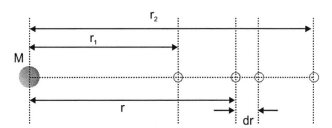

**Figure 7.19: Gravitational potential difference in a gravitational field due to a point.**

$$\Rightarrow \Delta V = -\int_{r1}^{r2} \frac{GMdr}{r^2}$$

$$\Rightarrow \Delta V = -\frac{GM}{r}$$

For applications of gravitational law, please visit:

http://cnx.org/content/m15085/latest/

# PowerPoint Link:

Please refer to the end of the module lecture link.

# Discussion Question

Answer the following questions with references. Please remember to follow the standard APA referencing style.

For APA standards of references, please visit: http://owl.english.purdue.edu/owl/resource/560/01/

Also, respond in detail to one other post by fellow students.

7.1  Discuss the function of a washing machine/dryer in terms of circular motions, centripetal force, and pseudo centrifugal force. Elaborate on the separation process of water from the wet cloths using physics knowledge.

7.2  Describe the differences between circular motions and elliptical motions. Explain how Newton's universal gravitational law can be applied to study the interplanetary motions.

# Laboratory Activity and the link

Go to the link below and run the simulation:

http://phet.colorado.edu/simulations/sims.php?sim=My_Solar_System

Go to: http://phet.colorado.edu/en/contributions/view/2921

Open Exploring Gravitation.doc

Do the activity.


Now take the chapter 7 test (not included with this book)

# Module 5

## CHAPTER-8

# Rotational Equilibrium and Rotational Dynamics

## Objectives

At the end of this lesson, you should be able to:

1. Answer questions on torque

2. Apply knowledge on rotational forces to solve problems

# Lecture Notes

## Rotational Kinematics

| 1D Linear Motion | Rotation about a Fixed Axis |
|---|---|
| **basic definitions** ||
| displacement x | angular displacement $\theta$ |
| velocity, $v = dx/dt$ | angular velocity $$\omega = \frac{d\theta}{dt}$$ |
| acceleration, $a = dv/dt$ | angular acceleration $$\alpha = \frac{d\omega}{dt}$$ |
| **Kinematic Equations** ||
| $v = v_i + a\,t$ | $\omega = \omega i + \alpha t$ |
| $x = x_i + v_i t + (1/2)\,a\,t^2$ | $\theta = \theta_i + \omega_i t + (\frac{1}{2})\alpha t^2$ |
| $v^2 = v_i{}^2 + 2\,a\,(x - x_i)$ | $\omega^2 = \omega_i{}_2 + 2\alpha(\theta - \theta_i)$ |

# Angular and Linear Quantities

## Tangential velocity, $v_t$

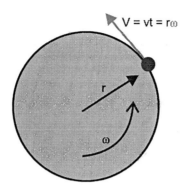

$$V = v_t = r\omega$$

## Tangential acceleration, $a_t$

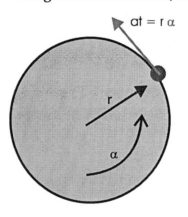

$$a_t = r\alpha$$

## Radial acceleration, $a_r$,
## or centripetal acceleration, $a_c$.

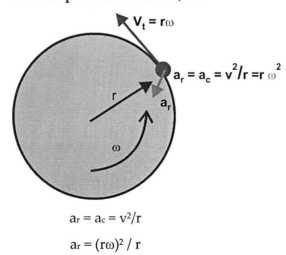

$$V_t = r\omega$$

$$a_r = a_c = v^2/r = r\omega^2$$

$$a_r = a_c = v^2/r$$

$$a_r = (r\omega)^2 / r$$

$$a_r = r\omega^2$$

The tangential acceleration $a_t$ and the radial acceleration $a_r$ are two vector **components** of the acceleration;

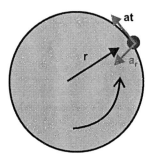

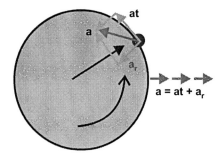

These are **vector** components and must be added **as vectors!**

# Rotational Energy

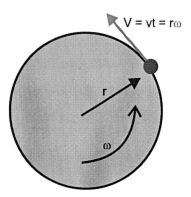

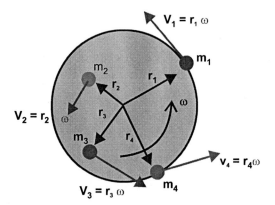

$KE_{Tot} = \Sigma KE_i$

$KE_{Tot} = \Sigma \; (^1/_2) \; m_i \; v_i^2$

$KE_{Tot} = \Sigma \; (^1/_2) \; m_i \; (v_i)^2$

$KE_{Tot} = \Sigma \; (^1/_2) \; m_i \; (r_i \; \omega)^2$

$KE_{Tot} = (^1/_2) \; \Sigma \; [ \; m_i \; r_i^2] \; \omega^2$

$KE_{Tot} = (^1/_2) \; [\text{"rotational mass"}] \; \omega^2$

$KE_{Tot} = (^1/_2) \; [\text{"moment of inertia"}] \; \omega^2$

$I = \Sigma \; [ \; m_i \; r_i^2] = \text{moment of inertia}$

$KE_{Tot} = (^1/_2) \; I \; \omega^2$

$I = \Sigma \; [ \; m_i \; r_i^2]$

$r_i$ is the perpendicular distance from the axis of rotation.

## Moment of Inertia

$KE_{Tot} = (^1/_2) \; I \; \omega^2$

$I = \Sigma \; [ \; m_i \; r_i^2]$

$r_i$ is the **perpendicular** distance from the axis of rotation.

For an **extended body** or a **continuous body**, this becomes

$I = \Sigma \; [ \; m_i \; r_i^2]$

$I = \Sigma \; [ \; r_i^2 \; \Delta m_i]$

$I = \lim \Sigma \, r_{i\,2} \Delta m_i = \int r^2 dm$

**r** is still the **perpendicular** distance from the axis of rotation.

$dm = \rho dv$

$I = \int r^2 \rho dv$

Normally, the density $\rho$ will be a constant so we will have

$I = \rho \int r^2 dv$

## Torque

Torque about a point is a concept that denotes the tendency of force to turn or rotate an object in motion. This tendency is measured in general about a point. It is also termed as "moment of force". The torque in angular motion corresponds to force in translation. It is the "cause" whose effect is either angular acceleration or angular deceleration of a particle in general motion. Quantitatively, it is defined as a vector given by:

$\tau = r \times F$

Rotation is a special case of angular motion. In the case of rotation, torque is defined with respect to an axis such that vector "$r$" is constrained to be perpendicular to the axis of rotation. In other words, the plane of motion is perpendicular to the axis of rotation. Clearly, the torque in rotation corresponds to force in translation.

## Torque about a point

An external force on a particle constitutes a torque with respect to a point. The only condition is that the point, about which torque is defined or measured, does not lie on the line of force, in which case torque is zero.

**Torque on a particle**

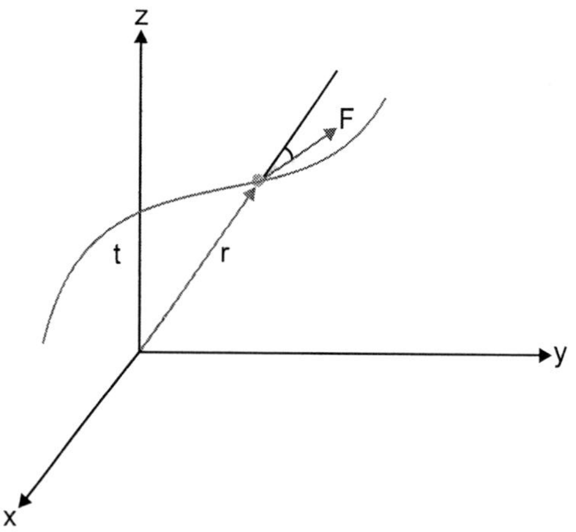

**Figure 8.1: The particle moves in three dimensional coordinate space.**

We can make note of the fact that it is convenient to construct the reference system in such a manner so that the point (about which torque is measured) coincides with the origin. In that case, the vector ($r$) denoting the position of the particle with respect to the point, becomes the position vector, which is measured from the origin of reference system.

## Magnitude of torque

With the reference of origin for measuring torque, we can find the magnitude of torque, using any of the following relations given below. Here, we have purposely considered force in xy - plane for illustration and visualization purposes as it provides clear directional relationship of torque "$\tau$" with the operand vectors "$r$" and "$F$".

**Torque on a particle**

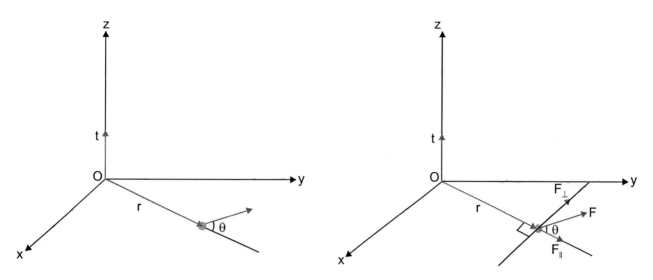

**(a)** Torque in terms of angle enclosed vector.

**(b)** Torque in terms of force, perpendicular to position

**Figure 8.2**

1:    Torque in terms of angle enclosed

$\tau = r\,F\,\sin\theta$

2:    Torque in terms of force perpendicular to position vector

$\tau = r(F\sin\theta) = rF_\perp$

3:    Torque in terms of moment arm

$\tau = (r\sin\theta)F = r_\perp F$

**Torque on a particle**

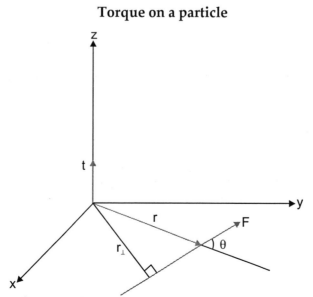

**Figure 8.3: Torque in terms of moment arm**

# Example 1

*Problem* : A projectile of mass "m" is projected with a speed "v" at an angle "θ" within the horizontal. Calculate the torque on the particle at the maximum height in relation to the point of projection.

*Solution* : The magnitude of the torque is given by :

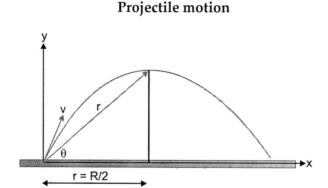

**Projectile motion**

**Figure 8.4: Torque on a projectile.**

$\tau = r_\perp F$=product of moment arm and magnitude of force

where $r\perp$ is moment arm. In this case, the moment arm is equal to half of the horizontal range of the flight,

$$r_\perp = \frac{R}{2} = \frac{v2sin2\theta}{2g}$$

Now, the force on the projectile of mass "m" is due to the force of gravity,

$$F = mg$$

Putting these expressions of moment arm and force in the expression of torque, we have,

$$\Rightarrow \tau = \frac{mgv2sin2\theta}{2}$$

The application of the right hand rule indicates that torque is clockwise and is directed in to the page.

## Direction of torque

The determination of torque's direction is relatively easier than that of angular velocity. The reason is simple. The torque itself is equal to the vector product of two vectors, unlike angular velocity which is one of the two operands of the vector product. Clearly, if we know the directions of two operands here, the direction of torque can easily be interpreted.

By the definition of vector product, the torque is perpendicular to the plane formed by the position vector and force. Besides, it is also perpendicular to each of the two vectors individually. However, the vector relation by itself does not tell which side of the plane formed by operands is the direction of torque. In order to decide the orientation of the torque, we employ the right hand vector product rule.

For this, we need to shift one of the operand vectors such that their tails meet at a point. It is convenient to shift the force vector, because application of the right hand vector multiplication rule at the point (origin of the coordinate system) gives us the sense of angular direction of the torque.

**Vector product**

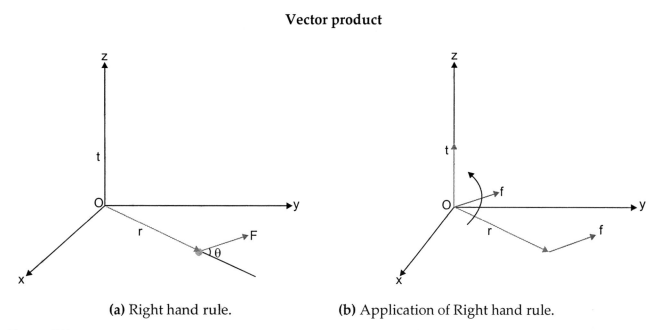

(a) Right hand rule.          (b) Application of Right hand rule.

**Figure 8.5**

In the figure, we shift the second vector (F) so that tails of two operand vectors meet at the point (origin) about which torque is calculated. The two operand vectors define a plane ("xy" - plane in the figure). The torque ($\tau$) is then acting along a line perpendicular to this plane and passing through the meeting point (origin).

The orientation of the torque vector in either of two directions is determined by applying the right hand rule. For this, we sweep the closed fingers of the right hand from the position vector (first vector) to the force vector (second vector). The outstretched thumb, then, indicates the orientation of torque. In the case shown, the direction of torque is the positive z-direction.

While interpreting the vector product, we must be careful about the sequence of operand. The vector product " *Fxr* " is negative or opposite in sign to that of " *rxF* ".

# Illustration

The problem involving torque is about evaluating the vector product. We have the options of using any of the three methods described above for calculating the magnitude of torque. In this section, we shall work with two important aspects of calculating torque:

- More than one force operating simultaneously on the particle
- Relative orientation of the plane of operand vectors and planes of the coordinate system

## More than one force operating simultaneously on the particle

Here, we have two choices. Either, we evaluate torques for each force and then find the resultant torque, or we evaluate the resultant (net) force first and then find the torque. The choice depends on the situation in hand.

In particular, if we find that the resulting torques are along the same direction (say one of the coordinate directions), then it is always better to calculate torque individually. This is because torques along a direction can be dealt with as scalars with an appropriate sign. The resultant torque is simply the algebraic sum. The example below highlights this aspect.

The bottom line in making a choice is that we should keep non-linear vector summation - whether that of force or torque - to a minimum. Otherwise, analytical technique for vector addition will need to be employed.

## Example 2

*Problem* : At an instant, a particle in the xy plane is acted on by two forces 5 N and 10 N in the xy plane as shown in the figure. Find the net torque on the particle at that instant.

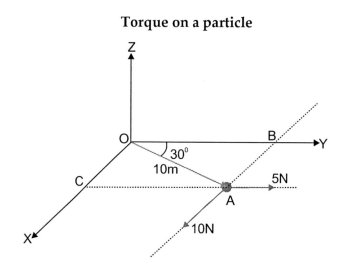

**Figure 8.6: Two forces are acting simultaneously on the particle.**

*Solution*: An inspection of the figure reveals that forces are parallel to axes and perpendicular linear distances of the lines of forces can be known. In this situation, calculation of the magnitude of torque is suited to the form of expression that uses moment arm. Here, moment arms for two forces are:

For 5 N force, the moment arm "AB" is,

$$AB = OA\sin 30^0 = 10 \times 0.5 = 5m$$

For 10 N force, the moment arm "AC" is,

$$AC = OA\cos 30^0 = 10 \times 0.866 = 8.66m$$

The torques due to 5 N is,

$$\Rightarrow \tau_1 = 5 \times 5 = 25 Nm \text{(positive being anti-clockwise)}$$

The torques due to 10 N is,

⇒$\tau_2$=−8.66x10=−86.6$Nm$(negative being clockwise)

Net torque is,

⇒$\tau_{net}$=$\tau_1$+$\tau_2$=25−86.6=−61.6$Nm$(negative being clockwise)

*Note*:

We can also first calculate the resultant force and then apply the definition of torque to obtain net torque. However, this would involve non-linear vector addition.

## Relative orientation of the plane of operand vectors

There are two possibilities here. The plane of operands are the same as that of the plane of the coordinate system. In simple words, the plane of velocity and position vector is the same as one of three planes formed by the coordinate system. If this is so, then the problem analysis is greatly simplified. Clearly, the torque will be along the remaining third axis. We have only to find the orientation (in the positive or negative direction of the axis) with the help of the right hand rule. It is evident that we should always strive to orient our coordinate system, if possible, so that the plane formed by the position vector and force aligns with one of the coordinate planes.

Should the two planes be different, then we need to be careful (i) in first identifying the plane of operands (ii) finding the relation of the operands' plane with the coordinate plane and (iii) applying the right hand rule to determine orientation (which side of the plane). Importantly, we cannot treat torque as scalar with an appropriate sign, but have to find the direction of torque with respect to one of the coordinates.

## Example 3

*Problem :* A particle in the xy plane is acted on by a force 5 N in the  z-direction as shown in the figure. Find the net torque on the particle.

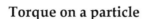

**Torque on a particle**

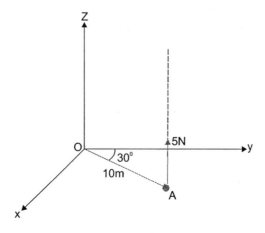

**Figure 8.7: Two force is acting parallel to z-axis.**

*Solution :* In this case, the moment arm is equal to the magnitude of the position vector. Hence, torque is:

$$\tau = 10 \times 5 = 50 Nm$$

However, finding its direction is not as straight forward. The situation here differs to earlier examples in one important respect. The plane containing the position vector and force vector is a plane defined by zOA. Also, this plane is perpendicular to the xy - plane. Now, torque is passing through the origin and is perpendicular to the plane zOA. We know from a theorem of geometry that angles between two lines and their perpendiculars are the same. Applying same and the right hand rule, we conclude that the torque is shifted by 30 degree from the y-axis and lies in the xy-plane as shown in the figure here.

**Torque on a particle**

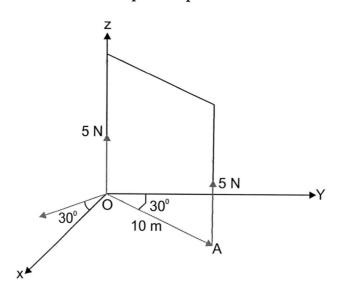

**Figure 8.8: The torque makes an angle of 30 with y-axis.**

## Torque in component form

Torque, being a vector, can be evaluated in component form with the help of unit vectors along the coordinate axes. The various expressions involved in the vector algebraic analysis are as given here:

$$\boldsymbol{\tau} = \boldsymbol{r} \times \boldsymbol{F}$$

$$\boldsymbol{\tau} = (x\boldsymbol{i} + y\boldsymbol{j} + z\boldsymbol{k}) \times (F_x\boldsymbol{i} + F_y\boldsymbol{j} + F_z\boldsymbol{k})$$

$$\tau = \begin{vmatrix} i & j & k \\ x & y & z \\ F_x & F_y & F_z \end{vmatrix}$$

$$\boldsymbol{\tau} = (yF_z - zF_y)\boldsymbol{i} + (zF_x - xF_z)\boldsymbol{j} + (xF_y - yF_x)\boldsymbol{k}$$

(5)

# Example 4

*Problem* : A force $F = 2i + j - 2k$ Newton acts on a particle at $i + 2j - k$ meters at a given time. Find the torque about the origin of coordinate system.

*Solution* : Hence, torque is :

$$\tau =$$

$$\begin{vmatrix} i & j & k \\ 1 & 2 & -1 \\ 2 & 1 & -2 \end{vmatrix}$$

$$\Rightarrow \tau = [(2 \times -2)-(1 \times -1)]i+[(-1 \times 2)-(1 \times -2)]j+[(1 \times 1)-(2 \times 2)]k$$

$$\Rightarrow \tau = -3i - 3k \, Nm$$

# Summary

1: The torque in angular motion corresponds to force in translation. It is the "cause" whose effect is either angular acceleration or angular deceleration of a particle in general motion. The expression of torque in both cases, however, is the same :

$$\tau = r \times F$$

2: When the point (about which torque is defined) coincides with the origin of the coordinate system, the vector "$r$" appearing in the expression of torque is the position vector.

3: Magnitude of torque is given by either of the following relations:

   (i)  Torque in terms of angle enclosed

   $$\tau = rF \sin\theta$$

   (ii) Torque in terms of force perpendicular to position vector

   $$\tau = rF_\perp$$

   (iii) Torque in terms of moment arm

   $$\tau = r_\perp F$$

4: Direction of torque

   The torque is perpendicular to the plane formed by the velocity vector and force, and also individually to either of them. We determine the orientation of the torque vector by applying the right hand rule.

5: When more than one force acts, then we should determine the resultant of forces or resultant of torques, depending on which of approach will minimize or avoid non-linear vector summation to obtain the result.

6: When the plane of operands is thsame as one of the coordinate planes, then torque acts along the remaining axis.

7:    Torque is expressed in component form as:

```
τ =
| i    j    k |
| x    y    z |
| Fx   Fy   Fz|
```

# Rotation

The study of rotational motion is an important step towards the study of real time motion. Rotational motion is essentially a circular motion about a point or an axis (for three dimensional bodies). The rotation of a rigid body about an axis passing through the body itself is termed "spin motion" or simply "spin". On the other hand, rotation about an external axis is termed "orbital motion". We shall treat both these types of rotation in the same manner, as treatment of either rotation type is same from the point of view of governing laws of motion.

Rotation is one of two basic components of general motion of rigid body. The motion of a rigid body is either translation or rotation, or a combination of the two. In translation, each of the particles constituting the rigid body has the same linear velocity and acceleration. In addition, we say that each of the particles constituting the rigid body has the same angular velocity and acceleration in rotation. A consequence of this analogy is that each particle constituting the rigid body moves in a circular motion such that their centers lie on a straight line, called the axis of rotation.

Pure translation refers to a straight-line motion. Similarly, pure rotation refers to rotation about a fixed axis. Importantly, pure translation excludes rotation, whereas pure rotation excludes translation.

# Torque in rotation

A rigid body in rotation about a fixed axis should continue to rotate with the given angular velocity indefinitely unless obstructed externally. The angular velocity of rotation is changed by external cause in the same manner as is translational motion. In the case of translation motion, the external cause is "force". We have to investigate what is the equivalent "cause" in the case of rotational motion.

Let us consider a rigid block placed over a smooth horizontal surface as shown in the figure below, which is subjected to a force across one of its faces. What is expected? The center of mass will move with linear acceleration following Newton's second law. However, the force is not passing through the center of mass. As such, the face of the block will also have turning tendency in the direction shown.

**The motion of block**

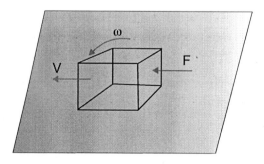

**Figure 8.9: The block has turning tendency**

Let us now imagine if the body is pierced through in the middle by a vertical bar with some clearance. The vertical bar will inhibit translation and the block will only rotate about the vertical bar. The question is what caused the block to turn around? Indeed, it is the force that caused the angular motion. However, it is not the only force that determined the outcome (magnitude and direction of angular velocity and acceleration). There are other considerations as well.

## Rotational motion

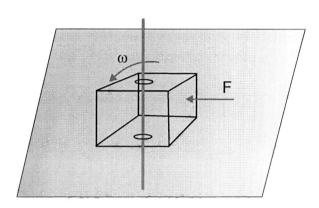

**Figure 8.10: The vertical bar inhibits translation and the block only rotates about the vertical bar.**

We all have the experience of opening and closing a hinged door in our house. It takes lesser effort (force) to open the door, when force is applied farther from the hinge. We can further reduce the effort by applying force normal to the plane of the door. On the other hand, we would require greater effort (force), if we push or pull the door from a point closer to the hinge. If we apply force in the radial direction (in the plane of door) towards the hinge, then the door does not rotate a bit - whatever be the magnitude of force. In a nutshell, rotation of the door depends on:

- Magnitude of the force

- Point of application of force with respect to the hinge (axis of rotation)

- Angle between the force and perpendicular line from the axis of rotation

These factors, which cause rotation, are captured by a quantity known as torque, which is defined as:

$\tau = r \times F$

where "$r$" is the position vector. There is a slight difficulty in interpreting this vector equation as applicable to rotation. The position vector,"$r$", as we can recall, is measured from a point. Now, the question is what is that point in the case of rotation about a fixed axis? Here, we observe that there is a unique point of application of the force. By the nature of motion, this point rotates in a plane perpendicular to the axis of rotation. We can, therefore, uniquely define the position of the point of application of force by measuring it from the point on the axis in that plane of rotation.

**Rotational motion**

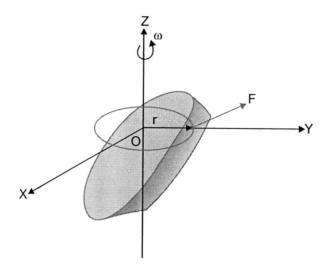

**Figure 8.11: Rotation about z-axis.**

The force may be directed in any possible direction. However, the body rotates about a fixed axis. It is not allowed to rotate or move about any other axes. This means that we are limited only to the component of torque along the axis of rotation. Therefore, we are only interested in components of force that lie in the plane of rotation, which is perpendicular to the axis of rotation. This we achieve by only considering the components of force which are perpendicular to the axis of rotation i.e. in the plane of rotation. In the figure below, we have tried to capture this aspect. We consider only the component of force in the xy-plane while evaluating the vector torque equation given above.

**Rotational motion**

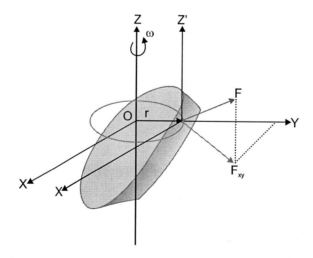

**Figure 8.12: Rotation about z-axis.**

In a nutshell, we interpret the vector product with two specific conditions:

- The position vector ($r$) is measured from a point on the axis, which is in the plane of rotation.
- The force vector ($F$) is the component of force in the plane of rotation.

The rotation of the rigid body about an axis passing through the body itself is a composition of very large numbers of circular motions by particles composing it. Each of the particles undergoes circular motion about the axis in a plane which is perpendicular to it.

## Magnitude of torque

With the understandings as deliberated above, let us determine the magnitude of torque. Here, let us consider that a force is applied at a point which is at a perpendicular distance "r" from the axis and in the plane of rotation. As far as application of force is concerned, it is a force applied at a point in the plane of rotation. In order to keep the context simplified, we have not drawn the rigid body assuming that the point of application rotates in a circle whose plane is perpendicular to the axis of rotation. Now, the magnitude of torque is given as:

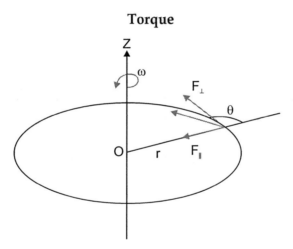

**Figure 8.13: Torque depends on the enclosed angle.**

$\tau = rF\sin\theta$

where "r","F" and "$\theta$" are quantities measured in the plane perpendicular to the axis of rotation.

Equivalently, we can determine the magnitude of torque as:

$\tau = r(F\sin\theta) = rF_\perp$

The magnitude of torque is equal to the product of the magnitude of the position vector as drawn from the center of the circle and the component of force perpendicular to the position vector. This is an important interpretation as it highlights that it is the tangential or perpendicular component (perpendicular to radial direction) of force, which is capable to produce change in the rotation. The component of force in the radial direction ($F \cos\theta$) does not affect rotation. We can also rearrange the expression of magnitude as:

$\tau = (r\sin\theta)F = r_\perp F$

The magnitude is equal to the product of the perpendicular distance and magnitude of the force. Perpendicular distance is obtained by drawing a perpendicular on the extended line of application of force as shown in the figure below. We may note here that this line is perpendicular to both the axis of rotation and force. This perpendicular distance is also known as the "moment arm" of the force and is denoted as "$r_\perp$".

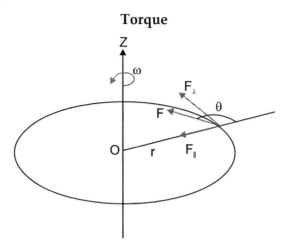

**Figure 8.14: The magnitude is equal to the product of moment arm and magnitude of the force.**

## Direction of torque

The nature of cross (or vector) product of two vectors, conveys a great deal about the direction of cross product i.e. torque where both position and force vectors are in the plane of rotation. It tells us that (i) the torque vector is perpendicular to the plane formed by operand vectors i.e. "$r$" and "$F$" and (ii) the torque vector is individually perpendicular to each of the operand vectors. Applying this explanation to the case in hand, we realize here that torque is perpendicular to the plane formed by the radius and force vectors i.e z-axis. However, we do not know which side of the plane i.e +z or -z direction, the torque is directed.

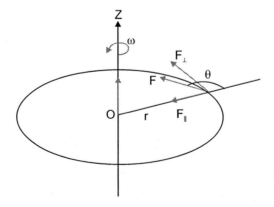

**Figure 8.15: Direction of torque is perpendicular to the plane of rotation.**

We apply the right hand rule to determine the remaining piece of information, regarding the direction of torque. We have two options here. Either we can shift the radius vector such that the tails of the two vectors meet at the position of the particle or we can shift the force vector (parallel shifting) so that the tails of the two vectors meet at the axis. The second approach has the advantage that the direction of the torque vector along the axis also gives the sense of rotation about that axis. Thus, following the second approach, we shift the force vector to the origin, while keeping the magnitude and direction the same as shown here:

**Direction of torque**

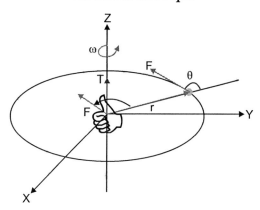

**Figure 8.16: Force vector is shifted to origin in order to apply right hand rule.**

Now, the direction of rotation is obtained by applying the rule of vector cross product. We place the right hand with closed fingers such that the curl of the fingers point in the direction as we transverse from the direction of the position vector (first vector) to the force vector (second vector). Then, the direction of the extended thumb points in the direction of torque. Alternatively, we see that a counter-clockwise torque is positive, whereas clockwise torque is negative. In this case, torque is counter-clockwise and is positive. Therefore, we conclude that torque is acting in +z-direction.

Pure rotation about a fixed axis gives us an incredible advantage in determining torque. We work with only two directions (positive and negative). In the case of torque about a point, however, we consider other directions of torque as well. It is also noteworthy to see that torque follows the superposition principle. Mathematically, we can add torque vectors algebraically as there are only two possible directions to obtain net or resultant torque. In words, it means that if a rigid body is subjected to more than one torque, then we can represent the torques by a single torque, which has the same effect on rotation.

Torque has the unit of Newton - meter.

## Rotational torque .vs. torque about a point

We have seen that rotational torque is calculated with the component of force which lies in the plane of rotation. This is required as the rigid body is free to move only about the fixed axis perpendicular to the plane of rotation. Accordingly, we evaluate the vector expression of torque by measuring the position vector from the center of the circle and using the component of force in the plane of rotation. However, if we evaluate the vector expression of torque $r \times F$ with respect to the origin of reference lying on the axis of rotation as is the general case of torque about a point, then we get torque which needs not be along the axial direction.

Clearly, rotational torque is a subset of torque defined for a point on the axis of rotation. This fact gives an alternate technique to determine rotational torque. We can approach the problem of determining rotational torque as described in the module, wherein we first resolve the given force in the plane of rotation containing the point of application. Then, we find the moment arm or perpendicular force component and determine the rotational torque as already explained. Alternatively, we can determine the torque about a point on the axis using vector expression of torque and then consider only the component of torque in the direction of the axis of rotation.

In order to fully understand the two techniques, we shall work an example here to illustrate.

Let us consider a force $F = (2i + 2j - 3k)$ Newton which acts on a rigid body at a point $r = (i + j - k)$ meters as measured from the origin of reference. In order to determine the rotational torque, say about x-axis, as the axis of rotation, we first proceed by considering only the force in the plane of rotation. The figure below shows the components of force and their perpendicular distances from three axes.

We find the position of application of force by first identifying A (1,1) in the xy-plane, then we find the position of the particle, B(1,1,-1) by moving "-1" in the negative z-direction as in the figure below.

## Torque on the particle

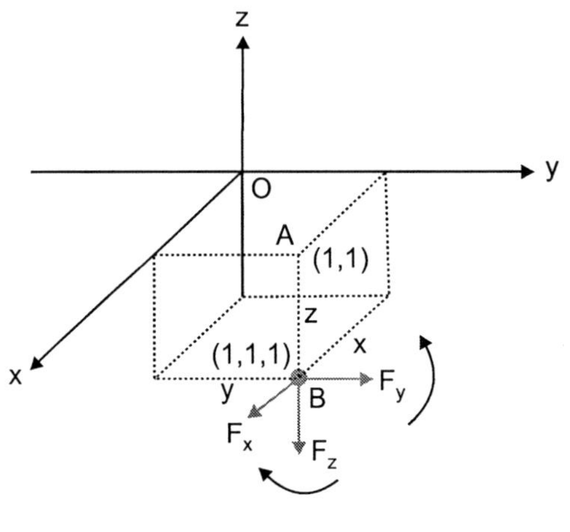

**Figure 8.17: The position of particle along with components of force**

Since the rigid body rotates about the x-axis, the plane of rotation is the yz plane. We, therefore, do not consider the component of force in the x-direction. The components of force relevant here are (i) the component of force in the y-direction Fy = 2 N at a perpendicular distance z = 1 m from the axis of rotation and (ii) the component of force in the z-direction Fz = 3 N at a perpendicular distance y = 1 m. For determining torque about the x-axis, we multiply perpendicular distance with force. We apply the appropriate sign, depending on the sense of rotation about the axis. The torque about the x-axis due to the component of force in the y - direction is counter-clockwise and hence is positive:

$\tau_1=zF_y=1\times2=2Nm$

The torque about the x-axis due to the component of force in the z - direction is clockwise and hence is negative:

$\tau_2=-yF_z=-1\times3=-3Nm$

Since both torques are acting along the same axis i.e. the x-axis, we can obtain net torque about the x-axis by algebraic sum:

$\Rightarrow\tau_x=\tau_1+\tau_2=2-3=-1Nm$

The net rotational torque is negative and hence clockwise in direction. The magnitude of net torque about the x-axis is:

$\Rightarrow\tau_x==1Nm$

Now, we proceed to calculate torque about the origin of reference, which lies on the axis of rotation. The torque about the origin of the coordinate system is given by:

$\tau=r\times F$

$$\tau = \begin{vmatrix} i & j & k \\ 1 & 1 & -1 \\ 2 & 2 & -3 \end{vmatrix}$$

$\Rightarrow\tau=[(1\times-3)-(2\times-1)]i+[(-1\times2)-(1\times-3)]j+[(1\times2\{-(1\times2)]k$

$\Rightarrow\tau=(-3+2)i+(-2+3)j$

$\Rightarrow\tau=-i+j$

The particle moves about the axis of rotation in the x-direction. In this case, the particle is restrained, and not allowed to rotate about any other axis. Thus, torque in rotation about the x-axis is equal to the x- component of torque about the origin. Now, the vector x-component of torque is:

$\tau_x=-i$

The net rotational torque is negative and hence clockwise in direction. The magnitude of torque about the x-axis for rotation is:

$\tau_x=1Nm$

## Nature of motion

Nature displays varieties of motion. Most of the time when we encounter motion, it is composed of different basic forms of motion. The most important challenge in the study of motion is to establish a clear understanding of the components (types) of motion that ultimately manifest in the world around us. Translational motion is the basic form. It represents the basic or inherent property of natural objects. A particle moving in a straight line keeps moving unless acted upon by an external force. However, what we see around is not what is fundamental to the matter (straight-line motion), but something which is grossly modified by the presence of force. This is the reason planets move around the Sun; electrons orbit about nucleus and airplanes circle the Earth on inter-continental flight.

Variety of motion is one important aspect of the study of motion. Another important aspect is the restrictive paradigm of fundamental laws that needs to be expanded to real time bodies and motions. For example, Newton's laws of motion as postulated are restricted to particle or particle-like bodies. The paradigm is required to be adapted to a system of particles and rigid body systems with the help of concepts like the center of mass. In real time, motion may be composed of even higher degrees of complexities. We can think of motions which involve rotation while translating. Now, the big question is whether Newtonian dynamics is capable to describe such composite motions? The answer is yes. However, it needs further development that addresses issues like the rotational dynamics of particles and rigid bodies.

# Summary

1.  The particles in a rigid body are locked and are placed at fixed linear distances from others.

2.  Each particle constituting a rigid body executes circular motion about a fixed axis in pure rotation.

3.  The cause of change in angular velocity in rotation is torque. It is defined as:

    $\tau = r \times F$

    where position vector,"$r$", and the force vector," $(F)$" are in the plane of rotation of the point of application of force, which is perpendicular to the axis of rotation.

4.  The magnitude of torque

    The magnitude of torque is determined in three equivalent ways :

    (i)   In terms of angle between position vector ($r$) and force ($F$)

    $\tau = rF\sin\theta$

    (ii)  In terms of tangential component of force ( $F_T$ or $F_\perp$ )

    $\tau = r(F\sin\theta) = rF_\perp$

    (iii) In terms of moment arm ( $r_\perp$ )

    $\tau = (r\sin\theta)F = r_\perp F$

5.  Direction of torque

    The vector equation of torque reveals that torque is perpendicular to the plane formed by position and force vectors and is also perpendicular to each of them individually. In order to know the sign of torque, we apply the right hand rule (positive for counter-clockwise and negative for clockwise rotation).

## Moment of Inertia (MI)

### Moment of inertia of known geometric bodies

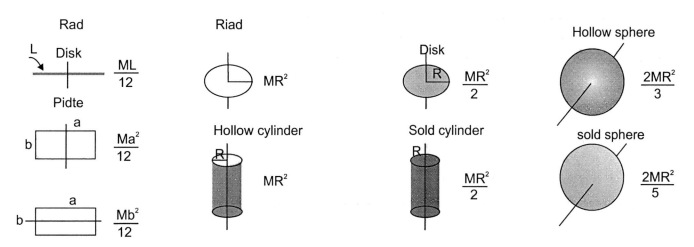

(a) **Moment of inertia of rod, plate, ring and hollow cylinders**

(b) **Moment of inertia of circular disk, solid cylinder, hollow sphere and solid sphere**

The radius of gyration (K) is defined as the radius of an equivalent ring of the same moment of inertia. The radius of gyration is defined for a given axis of rotation and has the unit of length. The moment of inertia of a rigid body about the axis of rotation, in terms of the radius of gyration, is expressed as:

$I=MK^2$

Some other MIs of interesting bodies are given here. These results can be derived as well in the same fashion when suitably applying the limits of integration.

(i)  The MI of a thick walled cylinder ($R_1$ and $R_2$ are the inner and outer radii)

$$I = \frac{M(R_1^2 + R_2^2)}{2}$$

(ii)  The MI of a thick walled hollow sphere ( $R_1$ and $R_2$ are the inner and outer radii)

$$I = \frac{2}{5} M \times \frac{R_2^5 - R_1^5}{R_2^3 - R_1^3}$$

For applications on rotational dynamics and moment of intertia, please visit:

http://cnx.org/content/m14292/latest/

# PowerPoint Link:

Please refer to the end of the module lecture link.

# Discussion Question

Answer the following questions with references. Please remember to follow the standard APA referencing style.

For APA standards of references, please visit: http://owl.english.purdue.edu/owl/resource/560/01/

Also, respond in detail to one other post by fellow students.

8.1 A man walking on a tightrope carries a long a pole, which has heavy items attached to the two ends. If he were to walk the tightrope with just a pole, what difference would it make to his balance? Discuss this in terms of angular rotation, angular momentum and moment of inertia.

8.2 Describe the applicability of torque to keep a ladder from sliding down. You need to describe the situation with values and derivations.

# Laboratory Activity and the link

Go to the link below and run the simulation:

http://phet.colorado.edu/simulations/sims.php?sim=My_Solar_System

Go to: http://phet.colorado.edu/en/contributions/view/3015

Open My Solar System Lab.doc

Do the activity.

# Module 5 Lecture Links:

http://ocw.mit.edu/courses/physics/8-01-physics-i-classical-mechanics-fall-1999/video-lectures/lecture-5/

http://ocw.mit.edu/courses/physics/8-01-physics-i-classical-mechanics-fall-1999/video-lectures/lecture-19/

http://ocw.mit.edu/courses/physics/8-01-physics-i-classical-mechanics-fall-1999/video-lectures/lecture-20/

http://ocw.mit.edu/courses/physics/8-01-physics-i-classical-mechanics-fall-1999/video-lectures/lecture-21/

http://ocw.mit.edu/courses/physics/8-01-physics-i-classical-mechanics-fall-1999/video-lectures/lecture-22

http://ocw.mit.edu/courses/physics/8-01-physics-i-classical-mechanics-fall-1999/video-lectures/lecture-24/

http://ocw.mit.edu/courses/physics/8-01-physics-i-classical-mechanics-fall-1999/video-lectures/lecture-25/

# Module 5 - Student's self-assessment

Please answer the following questions, which give you an indication of your standard of learning for this module.

- Could you describe the angular speed?
- Could you describe angular velocity?
- What is angular acceleration?
- Can you apply knowledge to solve problems involving circular motions?
- Can you solve problems on Torque?
- Can you apply your knowledge on rotational forces to various activities?
- Did you enjoy your lesson?
- What aspect of the lesson was most interesting to you?
- Can you now confidently do the lesson's activities?

Now take the chapter 8 test (not included with this book)

**Mid-term Test (Chapters 1 – 7)**

You may now take the mid-term test. The test is not included with this book. You should review all the previous chapter questions, prior to sitting the test.

# Module 6

## CHAPTER-9

# Solids and Fluids

## Objectives

At the end of this lesson, you should be able to:

1. Answer questions on fluid mechanics
2. Apply knowledge on fluid mechanics to solve problems

## Lecture Notes

### Fluids: Pressure, Density, Archimedes' Principle

One mistake you see in solutions to submerged-object static fluid problems is the inclusion, in the free body diagram for the problem, in addition to the buoyant force, of a pressure-times-area force, typically expressed as FP = PA. This is double counting. Students that include such a force, in addition to the buoyant force, do not realize that the buoyant force is the net sum of all the pressure-times-area forces exerted on the submerged object by the fluid in which it is submerged.

Gases and liquids are fluids. Unlike solids, they flow. A fluid is a liquid or a gas.

### Pressure

A fluid exerts pressure on the surface of any substance with which the fluid is in contact. Pressure is force-per-area. In the case of a fluid in contact with a flat surface over which the pressure of the fluid is constant, the magnitude of the force on that surface is the pressure times the area of the surface. Pressure has units of N/m2.

Never say that pressure is the amount of force exerted on a certain amount of area. Pressure is not an amount of force. Even in the special case in which the pressure over the "certain amount of area" is constant, the pressure is not the amount of force. In such a case, the pressure is what you have to multiply the area by to determine the amount of force.

The fact that the pressure in a fluid is 5 N/m2 in no way implies that there is a force of 5 N acting on a square meter of surface (any more than the fact that the speedometer in your car reads 35 mph implies that you are traveling 35 miles or that you have been traveling for an hour). In fact, if you say that the pressure at a particular point underwater in a swimming pool is 15,000 N/m$^2$ (fifteen thousand newtons per square meter), you are not specifying any area whatsoever. What you are saying is that any infinitesimal surface element that may be exposed to the fluid at that point will experience an infinitesimal force of magnitude dF that is equal to 15,000 N/m$^2$ times the area dA of the surface. When we specify a pressure, we are talking about a would-be effect on a would-be surface element.

We talk about an infinitesimal area element because it is entirely possible that the pressure varies with position. If the pressure at one point in a liquid is 15,000 N/m$^2$ it could very well be 16,000 N/m$^2$ at a point that is less than a millimeter away in one direction and 14,000 N/m$^2$ at a point that is less than a millimeter away in another direction.

Let us talk about direction. Pressure itself has no direction. However, the force that a fluid exerts on a surface element, because of the pressure of the fluid, does have direction. The force is perpendicular to, and toward, the surface. The direction of the force resulting from some pressure (let us call that the pressure-times-area force) on a surface element is determined by the victim (the surface element) rather than the agent (the fluid).

## Pressure Dependence on Depth

For a fluid near the surface of the earth, the pressure in the fluid increases with depth. You may have noticed this, if you have ever gone deep under water, because you can feel the effect of the pressure on your eardrums. Before we investigate this phenomenon in depth, it should be pointed out that in the case of a gas, this pressure dependence on depth is, for many practical purposes, negligible. In discussing a container of a gas for instance, we typically state a single value for the pressure of the gas in the container, neglecting the fact that the pressure is greater at the bottom of the container. We neglect this fact because the difference in the pressure at the bottom and the pressure at the top is so small compared to the pressure itself at the top. We do this when the pressure difference is too small to be relevant, but it should be noted that even a very small pressure difference can be significant. For instance, a helium-filled balloon, released from rest near the surface of the earth would fall to the ground if it weren't for the fact that the air pressure in the vicinity of the lower part of the balloon is greater (albeit only slightly greater) than the air pressure in the vicinity of the upper part of the balloon.

Let us do a thought experiment (Einstein was fond of thought experiments). These are also called Gedanken experiments. Gedanken is the German word for thought.) Imagine that we construct a pressure gauge as follows: We cap one end of a piece of thin pipe and put a spring completely inside the pipe with one end in contact with the end cap. Now we put a disk whose diameter is equal to the inside diameter of the pipe, in the pipe and bring it into contact with the other end of the spring. We grease the inside walls of the pipe so that the disk can slide freely along the length of the pipe, but we make the fit exact so that no fluid can get past the disk. Now we drill a hole in the end cap, remove all the air from the region of the pipe between the disk and the end cap, and seal up the hole. The position of the disk in the pipe, relative to its position when the spring is neither stretched nor compressed, is directly proportional to the pressure on the outer surface, the side facing away from the spring, of the disk. We calibrate (mark a scale on) the pressure gauge that we have just manufactured, and use it to investigate the pressure in the water of a swimming pool. First we note that, as soon as we removed the air, the gauge started to indicate a significant pressure (around $1.013\times10^5 \text{N/m}^2$), namely the air pressure in the atmosphere. Now we move the gauge around and watch the gauge reading. Wherever we put the gauge (we define the location of the gauge to be the position of the center point on the outer surface of the disk) on the surface of the water, we get one and the same reading, (the air pressure reading). Next we verify that the pressure reading does indeed increase as we lower the gauge deeper and deeper into the water. Then we find the point for which this paragraph was written: if we move the gauge around horizontally at one particular depth, the pressure reading does not change. That is the experimental result we want to use in the following development, the experimental fact that the pressure has one and the same value at all points that are at one and the same depth in a fluid.

Here we derive a formula that gives the pressure in an incompressible static fluid as a function of the depth in the fluid. Let us get back into the swimming pool. Now imagine a closed surface enclosing a volume, a region in space, that is full of water. Let us first call the water in such a volume, "a volume of water," and let us then consider giving it another name. If it were ice, it would be called a chunk of ice. However, since it is liquid water, let us call it a "slug" of water. We are going to derive the pressure versus depth relation by investigating the equilibrium of an "object" which is a slug of water.

Consider a cylindrical slug of water whose top is part of the surface of the swimming pool and whose bottom is at some arbitrary depth $h$ below the surface. Let us draw the slug here, isolated from its surroundings. The slug itself is, of course, surrounded by the rest of the water in the pool.

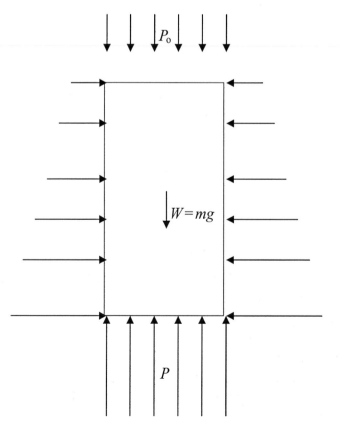

**Figure 9.1**

In the diagram, we use arrows to convey the fact that there is pressure-times-area force on every element of the surface of the slug. Now the downward pressure-times-area force on the top of the slug is easy to express in terms of the pressure because the pressure on every infinitesimal area element making up the top of the slug has one and the same value. In terms of the determination of the pressure-times-area, this is the easy case. The magnitude of the force, $F_o$, is just the pressure $P_o$ times the area A of the top of the cylinder.

$F_o = P_o A$

A similar argument can be made for the bottom of the cylinder. All points on the bottom of the cylinder are at the same depth in the water so all points are at one and the same pressure P. The bottom of the cylinder has the same area A as the top so the magnitude of the upward force F on the bottom of the cylinder is given by:

$F = PA$

As to the sides, if we divide the sidewalls of the cylinder up into an infinite set of equal-sized infinitesimal area elements, for every sidewall area element, there is a corresponding area element on the opposite side of the cylinder. The pressure is the same on both elements because they are at the same depth. The two forces then have the same magnitude, but because the elements face in opposite directions, the forces have opposite directions. Two opposite but equal forces add up to zero. In such a manner, all the forces on the sidewall area elements cancel each other out.

Now we are in a position to draw a free body diagram of the cylindrical slug of water.

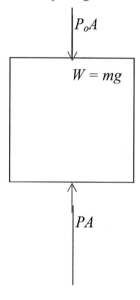

**Figure 9.2**

Applying the equilibrium condition yields,

$$\sum F_\uparrow = 0$$

$PA - mg - P_o A = 0$ $\hspace{4cm}$ *(1)*

At this point in our derivation of the relation between pressure and depth, the depth does not explicitly appear in the equation. The mass of the slug of water, however, does depend on the length of the slug, which is indeed the depth $h$. First, we note that

$m = \rho V$

where $\rho$ is the density, the mass-per-volume, of the water making up the slug and V is the volume of the slug. The volume of a cylinder is its height times its face area so we can write:

$m = \rho h A$

Substituting this expression for the mass of the slug into equation 1 yields:

$PA - \rho h A g - P_o A = 0$

$P - \rho h g - P_o = 0$

$P = P_o + \rho g h$   3

While we have been writing specifically about water, the only thing in the analysis that depends on the identity of the incompressible fluid is the density $\rho$. Hence, as long as we use the density of the fluid in question, equation 3, $(P = P_o + \rho g h)$ applies to any incompressible fluid. It says that the pressure at any depth $h$ is the pressure at the surface plus $\rho g h$.

A few words on the units of pressure are in order. We have stated that the units of pressure are N/m². This combination of units is given a name. It is called the *pascal*, abbreviated Pa.

$$1 \text{ Pa} = 1 \frac{N}{m^2}$$

Pressures are often quoted in terms of the non-SI unit of pressure, the *atmosphere*, abbreviated atm and defined such that, on the average, the pressure of the earth's atmosphere at sea level is 1 atm. In terms of the pascal,

1 atm = $1.013 \times 10^5$ Pa

A common mistake made when applying the equation $(P = P_o + \rho g h)$ is to ignore the units. Often, the use of 1 atm for $P_o$, without converting that to pascals, and the addition of the product $\rho g h$ to it can create a mismatch of units. Of course, if one uses SI units for $\rho$, $g$, and $h$, the product $\rho g h$ comes out in N/m² which is a pascal, which is definitely not an atmosphere (but rather, about a hundred-thousandth of an atmosphere). Of course, one cannot add a value in pascals to a value in atmospheres. The proper path forward is to convert the value of $P_o$, that was given in units of atmospheres, to pascals, and then add the product $\rho g h$ (in SI units) to the result so that the final answer comes out in pascals.

## Gauge Pressure

Consider the gauge we constructed for our thought experiment. The part about evacuating the inside of the pipe presents quite the manufacturing challenge. The gauge would become inaccurate as air leaked in by the disk. As regards function, the description is fairly realistic in terms of actual pressure gauges in use, except for the pumping of the air out the pipe. To make it more like an actual gauge that one might purchase, we would have to leave the interior open to the atmosphere. In use then, the gauge reads zero when the pressure on the sensor end is 1 atmosphere, and in general, indicates the amount by which the pressure being measured exceeds atmospheric pressure. This quantity, the amount by which a pressure exceeds atmospheric pressure, is called *gauge pressure* (since it is the value registered by a typical pressure gauge.) When it needs to be contrasted with gauge pressure, the actual pressure that we have been discussing up to this point is called *absolute pressure*.

The absolute pressure and the gauge pressure are related by:

$$P = P_G + P_O \tag{4}$$

where:

P is the absolute pressure,

$P_G$ is the gauge pressure, and

$P_O$ is atmospheric pressure.

When you hear a value of pressure (other than the so-called barometric pressure of the earth's atmosphere) in your everyday life, it is typically a gauge pressure (even though one does not use the adjective "gauge" in discussing it.) For instance, if you hear that the recommended tire pressure for your tires is 32 psi (pounds per square inch) what is being quoted is a gauge pressure. Folks that work on ventilation systems often speak of negative air pressure. Again, they are actually talking about gauge pressure, and a negative value of gauge pressure in a ventilation line just means that the absolute pressure is less than atmospheric pressure.

## Archimedes' Principle

The net pressure-times-area force on an object submerged in a fluid, the vector sum of the forces on all the infinite number of infinitesimal surface area elements making up the surface of an object, is *upward* because of the fact that pressure increases with depth. The upward pressure-times-area force on the bottom of an object is greater than the downward pressure-times-area force on the top of the object. The result is a net upward force on any object that is either partly or totally submerged in a fluid. The force is called the buoyant force on the object. The agent of the buoyant force is the fluid.

If you take an object in your hand, submerge the object in still water, and release the object from rest, one of three things will happen: The object will experience an upward acceleration and bob to the surface, the object will remain at rest, or the object will experience a downward acceleration and sink. We have emphasized that the buoyant force is always upward. So why on earth would the object ever sink? The reason is, of course, that after you release the object, the buoyant force is not the only force acting on the object. The gravitational force still acts on the object when the object is submerged. Recall that the earth's gravitational field permeates everything. For an object that is touching nothing of substance but the fluid it is in, the free body diagram (without the acceleration vector being included) is always the same (except for the relative lengths of the arrows):

**Figure 9.3**

and the whole question as to whether the object (released from rest in the fluid) sinks, stays put, or bobs to the surface, is determined by how the magnitude of the buoyant force compares with that of the gravitational force. If the buoyant force is greater, the net force is upward and the object bobs toward the surface; if the buoyant force and the gravitational force are equal in magnitude, the object stays put; and if the gravitational force is greater, the object sinks.

So how does one determine how big the buoyant force on an object is? First, the trivial case: If the only forces on the object are the buoyant force and the gravitational force, and the object remains at rest, then the buoyant force must be equal in magnitude to the gravitational force. This is the case for an object such as a boat or a log which is floating on the surface of the fluid it is in.

However, suppose the object is not freely floating at rest. Consider an object that is submerged in a fluid. We have no information on the acceleration of the object, but we cannot assume it to be zero. Assume that a person has, while maintaining a firm grasp on the object, submerged the object in fluid, and then, released it from rest. We do not know which way it is going from there, but we cannot assume that it is going to stay put.

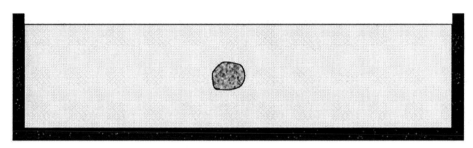

**Figure 9.4**

To derive our expression for the buoyant force, we do a little thought experiment. Imagine replacing the object with a slug of fluid (the same kind of fluid as that in which the object is submerged), where the slug of fluid has the exact same size and shape as the object.

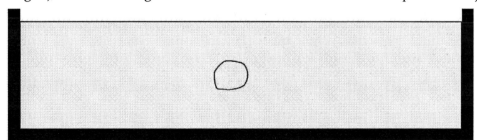

**Figure 9.5**

From our experience with still water we know that the slug of fluid would indeed stay put, meaning that it is in equilibrium.

| Table of Forces | | | |
|---|---|---|---|
| **Symbol=?** | **Name** | **Agent** | **Victim** |
| $B$ | Buoyant Force | The Surrounding Fluid | The Slug of Fluid |
| $F_{gSF} = m_{SF}\, g$ | Gravitational Force on the Slug of Fluid | The Earth's Gravitational Field | The Slug of Fluid |

Applying the equilibrium equation $\sum F_\uparrow = 0$ to the slug of fluid yields:

$$\sum F_\uparrow = 0$$

$a = 0$

gSF

$$B - F_{gSF} = 0$$

$$B = F_{gSF}$$

The last equation states that the buoyant force on the slug of fluid is equal to the gravitational force on the slug of fluid. Here is an important point that is the crux of the derivation: because the slug of fluid has the exact same size and shape as the original object, it presents the exact same surface to the surrounding fluid, and hence, the surrounding fluid exerts the same buoyant force on the slug of fluid as it does on the original object. Since the buoyant force on the slug of fluid is equal in magnitude to the gravitational force acting on the slug of fluid, the buoyant force on the original object is equal in magnitude to the gravitational force acting on the slug of fluid. This is Archimedes' principle.

$B$ = the buoyant force, which is equal in magnitude to the gravitational force that would be acting on that amount of fluid that would fit in the space occupied by the submerged part of the object.

$F_{go} = m_o\, g$

(The gravitational force)

Archimedes' Principle states that: The buoyant force on an object that is either partly or totally submerged in a fluid is upward, and is equal in magnitude to the gravitational force that would be acting on that amount of fluid that would be where the object is if the object was not there. For an object that is totally submerged, the volume of that amount of fluid that would be where the object is if the object was not there is equal to the volume of the object itself. However, for an object that is only partly submerged, the volume of that amount of fluid that would be where the object is if the object was not there is equal to the (typically unknown) volume of the submerged part of the object. In addition, if the object is freely floating at rest, the equilibrium equation (instead of Archimedes'

Principle) can be used to quickly establish that the buoyant force (of a freely floating object such as a boat) is equal in magnitude to the gravitational force acting on the object itself.

# Pascal's Principle, the Continuity Equation, and Bernoulli's Principle

There are two mistakes that tend to arise with some regularity in the application of the Bernoulli equation $P + \frac{1}{2}\rho\, v^2 + \rho gh = \text{constant}$ . Firstly, an occasionally overlooked item is the creation of a diagram in order to identify point 1 and point 2 in the diagram so that they can write the Bernoulli equation in its useful form: $P_1 + \frac{1}{2}\rho v_1^2 + \rho gh_1 = P_2 + \frac{1}{2}\rho v_2^2 + \rho gh_2$ . Secondly, when both the velocities in Bernoulli's equation are unknown, it is sometimes forgotten that there is another equation that relates the velocities, namely, the continuity equation in the form $A_1 v_1 = A_2 v_2$ which states that the flow rate at position 1 is equal to the flow rate at position 2.

# Pascal's Principle

Experimentally, we find that if you increase the pressure by some given amount at one location in a fluid, the pressure increases by that same amount everywhere in the fluid. This experimental result is known as Pascal's Principle.

We take advantage of Pascal's principle every time we step on the brakes of our cars and trucks. The brake system is a hydraulic system. The fluid is oil that is called hydraulic fluid. When you depress the brake pedal you increase the pressure everywhere in the fluid in the hydraulic line. At the wheels, the increased pressure acting on pistons attached to the brake pads pushes them against disks or drums connected to the wheels.

# Example 1

A simple hydraulic lift consists of two pistons, one larger than the other, in cylinders connected by a pipe. The cylinders and pipe are filled with water. In use, a person pushes down upon the smaller piston and the water pushes upward on the larger piston. The diameter of the smaller piston is 2.20 centimeters. The diameter of the larger piston is 21.0 centimeters. On top of the larger piston is a metal support and on top of that is a car. The combined mass of the support-plus-car is 998 kg. Find the force that the person must exert on the smaller piston to raise the car at a constant velocity. Neglect the masses of the pistons.

*Solution*

We start our solution with a sketch:

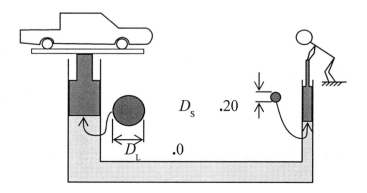

Now, let us find the force R exerted on the larger piston by the car support. By Newton's third law, it is the same as the normal force FN exerted by the larger piston on the car support. We will draw and analyze the free body diagram of the car plus the support to get that,

| Table of Forces | | | |
|---|---|---|---|
| Symbol | Name | Agent | Victim |
| $F_\gamma = m\gamma$ | Gravitational Force on Support-Plus-Car | The Earth's Gravitational Field | The Support-Plus-Car |
| $F_N$ | Normal Force | The Large Piston | The Support-Plus-Car |

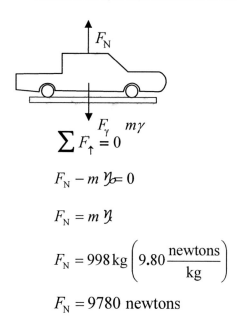

$$\sum F_\uparrow = 0$$

$$F_N - m\,\gamma = 0$$

$$F_N = m\,\gamma$$

$$F_N = 998\,\text{kg}\left(9.80\,\frac{\text{newtons}}{\text{kg}}\right)$$

$$F_N = 9780\ \text{newtons}$$

Now we analyze the equilibrium of the larger piston to determine what the pressure in the fluid must be in order for the fluid to exert enough force on the piston (with the car-plus-support on it) to keep it moving at constant velocity.

| Table of Forces | | | |
|---|---|---|---|
| Symbol | Name | Agent | Victim |
| R = F<br>= 9780 newtons | Interaction Partner to Normal Force (see above) | Support (That part, of the hydraulic lift, that the car is on.) | Large Piston |
| F<br>PL | Pressure-Related Force on Large Piston | The Water | Large Piston |

$$R_\text{N} = F_\text{N}$$

$$\vec{a} = 0$$

$$F_\text{PL} = PA_\text{L}$$

$$\sum F_\uparrow = 0$$

$$F_\text{PL} - R_\text{N} = 0$$

$$PA_\text{L} - R_\text{N} = 0$$

$$P = \frac{R_\text{N}}{A_\text{I}}$$

$A_\text{L}$      We can use the given larger piston diameter $D_\text{L}$ to determine the area of the face of the larger piston as follows:

$$A_\text{L} = \pi \left(\frac{D_\text{L}}{2}\right)^2$$

$$A_\text{L} = \pi \left(\frac{0.210 \text{ m}}{2}\right)^2$$

$$A_\text{L} = 0.03464 \text{m}^2$$

$$A_\text{L} = \pi \vec{r}_\text{L}^{\,2}$$

where $\vec{r}_\text{L} = \dfrac{D_\text{L}}{2}$ is the radius of the larger piston.

Substituting this and the value $R_\text{N} = F_\text{N} = 9780$ newtons into equation **Error! Reference source not found.** above yields

$$P = \frac{9780 \text{ newtons}}{0.03464 \text{ m}^2}$$

$$P = 282\,333 \frac{\text{N}}{\text{m}^2} \quad \text{(We intentionally keep 3 too many significant figures in this } \textit{intermediate} \text{ result.)}$$

Now analyze the equilibrium of the smaller piston to determine the force that the person must exert

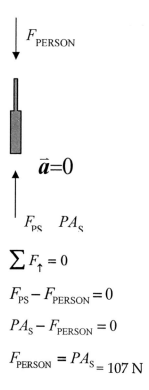

$$\sum F_{\uparrow} = 0$$

$$F_{PS} - F_{PERSON} = 0$$

$$PA_{S} - F_{PERSON} = 0$$

$$F_{PERSON} = PA_{S} = 107 \text{ N}$$

# Fluid in Motion—the Continuity Principle

The Continuity Principle is simply a statement of the fact that for any section of a single pipe, filled with an incompressible fluid (an idealization approached by liquids), through which the fluid with which the pipe is filled is flowing, the amount of fluid that goes in one end in any specified amount of time is equal to the amount that comes out the other end in the same amount of time. If we quantify the amount of fluid in terms of the mass, this is a statement of the conservation of mass. Having stipulated that the segment is filled with fluid, the incoming fluid has no room to expand in the segment. Having stipulated that the fluid is incompressible, the molecules making up the fluid cannot be packed closer together; that is, the density of the fluid cannot change. With these stipulations, the total mass of the fluid in the segment of pipe cannot change, so, for any time a certain mass of the fluid flows in one end of the segment, the same mass of the fluid must flow out the other end.

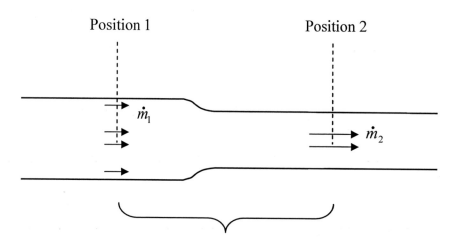

Segment of Pipe Filled with the Fluid that is Flowing Through the Pipe

This can only be the case if the mass flow rate, the number of kilograms-per-second passing a given position in the pipe, is the same at both ends of the pipe segment.

$$\dot{m}_1 = \dot{m}_2$$

An interesting consequence of the continuity principle is the fact that, in order for the mass flow rate (the number of kilograms per second passing a given position in the pipe) to be the same in a fat part of the pipe as it is in a skinny part of the pipe, the velocity of the fluid (i.e. the velocity of the molecules of the fluid) must be greater in the skinny part of the pipe. Let us determine why this is the case.

Here, we again depict a pipe in which an incompressible fluid is flowing.

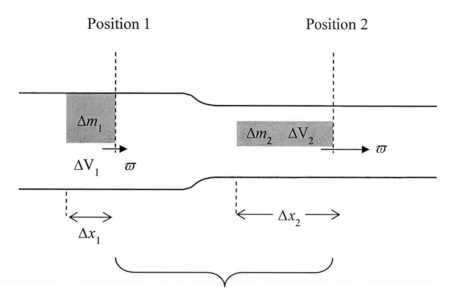

Segment of Pipe Between Positions 1 and 2

Keeping in mind that the entire pipe is filled with the fluid, the shaded region on the left represents the fluid that will flow past position 1 in time $\Delta t$ and the shaded region on the right represents the fluid that will flow past position 2 in the same time $\Delta t$. In both cases, in order for the entire slug of fluid to cross the relevant position line, the slug must travel a distance equal to its length. Now the slug labeled $\Delta m2$ has to be longer than the slug labeled $\Delta m1$ since the pipe is skinnier at position 2 and by the continuity equation $\Delta m1 = \Delta m2$ (the amount of fluid that flows into the segment of the pipe between position 1 and position 2 is equal to the amount of fluid that flows out of it). So, if the slug at position 2 is longer and it has to travel past the position line in the same amount of time as it takes for the slug at position 1 to travel past its position line, the fluid velocity at position 2 must be greater. The fluid velocity is greater at a skinnier position in the pipe.

Let us determine a quantitative relation between the velocity at position 1 and the velocity at position 2. Starting with

$\Delta m1 = \Delta m2$

we use the definition of density to replace each mass with the density of the fluid times the relevant volume:

$\varrho \Delta V_1 = \varrho \, \Delta V_2$ Dividing both sides by the density tells us something already known:

$\Delta V_1 = \Delta V_2$

As an aside, note that if you divide both sides by $\Delta t$ and take the limit as $\Delta t$ goes to zero, we have $V1 = V2$ which is an expression of the continuity principle in terms of volume flow rate. The volume flow rate is typically referred to simply as the flow rate.

While we use the SI units   m3/s   for flow rate, the reader may be more familiar with flow rate expressed in units of gallons per minute.

Returning to our determination of a mathematical relation between the velocities of the fluid at the two positions in the pipe, we copy the diagram of the pipe and add, to the copy, a depiction of the face of slug 1 of area A1 and the face of slug 2 of area A2

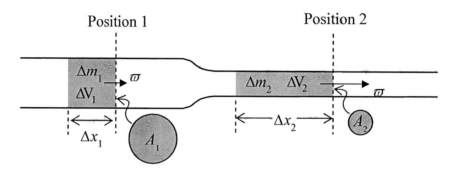

Remember that $\Delta V_1 = \Delta V_2$ . Each volume can be replaced with the area of the face of the corresponding slug, times the length of that slug. So,

$A_1 \Delta x_1 = A_2 \Delta x_2$

Recall that $\Delta x_1$ is not only the length of slug 1, it is also how far slug 1 must travel in order for the entire slug of fluid to get past the position 1 line. The same is true for slug 2 and position 2. Dividing both sides by the one time interval $\Delta t$ yields:

$$A_1 \frac{\Delta x_1}{\Delta t} = A_2 \frac{\Delta x_2}{\Delta t}$$

Taking the limit as $\Delta t$ goes to zero results in:

$A_1 v_1 = A_2 v_2$ 　　　　　　　　　　　　　　　　　　　(5)

This relation between the velocities is what is being sought. It applies to any pair of positions in a pipe completely filled with an incompressible fluid. It can be written as,

$A v$ = constant 　　　　　　　　　　　　　　　　　　　(6)

which means that the product of the cross-sectional area of the pipe and the velocity of the fluid at that cross section is the same for every position along the fluid-filled pipe. To take advantage of this fact, one typically writes, in equation form, that the product $Av$ at one location is equal to the same product at another location. In other words, one writes this equation as. Follows:

(Note that the expression $Av$ , the product of the cross-sectional area of the pipe, at a particular position, and the velocity of the fluid at that same position, having been derived by dividing an expression for the volume of fluid $\Delta V$ that would flow past a given position of the pipe in time $\Delta t$, by

Δt, and taking the limit as Δt goes to zero, is none other than the flow rate (the volume flow rate) discussed in the aside above.)

Flow Rate = A$v$

Further, note that if we multiply the flow rate by the density of the fluid, we get the mass flow rate.

m = ϱA$v$    (7)

# Fluid in Motion—Bernoulli's Principle

The derivation of Bernoulli's Equation represents an elegant application of the Work-Energy Theorem. Here we discuss the conditions under which Bernoulli's Equation applies and then simply state and discuss the result.

Bernoulli's Equation applies to a fluid flowing through a full pipe. The degree to which Bernoulli's Equation is accurate depends on the degree to which the following conditions are met:

1.  The fluid must be experiencing steady state flow. This means that the flow rate at all positions in the pipe is not changing with time.

2.  The fluid must be experiencing streamline flow. Pick any point in the fluid. The infinitesimal fluid element at that point, at an instant in time, traveled along a certain path to arrive at that point in the fluid. In the case of streamline flow, every infinitesimal element of fluid that ever finds itself at that same point traveled the same path. (Streamline flow is the opposite of turbulent flow.)

3.  The fluid must be non-viscous. This means that the fluid has no tendency to "stick to" either the sides of the pipe or to itself. (Molasses has high viscosity. Alcohol has low viscosity.)

Consider a pipe full of a fluid that is flowing through the pipe. In the most general case, the cross-sectional area of the pipe is not the same at all positions along the pipe and different parts of the pipe are at different elevations relative to an arbitrary, but fixed, reference level.

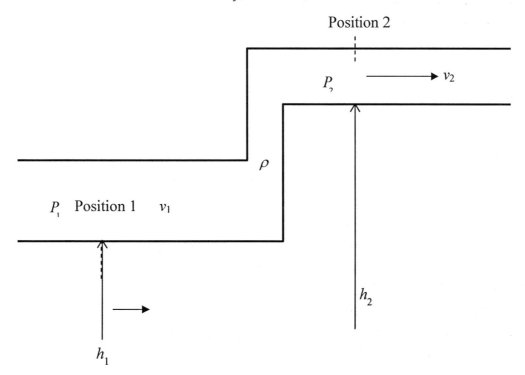

Pick any two positions along the pipe, e.g. positions 1 and 2 in the diagram above (it is already known that, in accord with the continuity principle, $A_1v_1 = A_2v_2$). Consider the following unnamed sum of terms:

$$P + \tfrac{1}{2}\rho v^2 + \rho gh$$

where, at the position under consideration:

P is the pressure of the fluid,

$\rho$ (the Greek letter rho) is the density of the fluid,

$v$ is the magnitude of the velocity of the fluid,

g = acceleration due to gravity

h is the elevation, relative to a fixed reference level, of the position in the pipe.

The Bernoulli Principle states that this unnamed sum of terms has the same value at each and every position along the pipe. Bernoulli's equation is typically written:

$$P + \tfrac{1}{2}\rho v^2 + \rho gh = \text{constant}$$

However, to use this, two positions along the pipe must be selected and an equation written, stating that the value of the unnamed sum of terms is the same at one of the positions as it is at the other.

$$P_1 + \tfrac{1}{2}\rho v_1^2 + \rho gh_1 = P_2 + \tfrac{1}{2}\rho v_2^2 + \rho gh_2$$

A particularly interesting characteristic of fluids is incorporated in this equation. Suppose positions 1 and 2 are at one and the same elevation, then $h_1 = h_2$ in equation 7 and it becomes:

$$P_1 + \tfrac{1}{2}\rho v_1^2 = P_2 + \tfrac{1}{2}\rho v_2^2$$

If $v_2 > v_1$ then $P_2$ must be less than $P_1$ in order for the equality to hold. This equation is saying that, where the velocity of the fluid is high, the pressure is low.

# PowerPoint Link:

Please refer to the end of the module lecture link.

# Discussion Question

Answer the following questions with references. Please remember to follow the standard APA referencing style.

For APA standards of references, please visit: http://owl.english.purdue.edu/owl/resource/560/01/

Also, respond in detail to one other post by fellow students.

9.1 Visit the following website:

http://www.cbsnews.com/stories/2006/11/28/eveningnews/main2213458.shtml\

Read about the current predicament of Dead Sea.

Discuss the features of Dead Sea in terms of buoyancy, density, and general fluid mechanics knowledge. If you were to reverse the properties of Dead Sea, what can you do?

9.2 Describe how a hydraulic lift works in terms of fluid mechanics and derive all applicable equations.

# Laboratory Activity and the link

Go to the link below and run the simulation:

http://phet.colorado.edu/simulations/sims.php?sim=Balloons_and_Buoyancy

Go to: http://phet.colorado.edu/en/contributions/view/3076

Open Introduction to Balloons and Buoyancy.doc

Do the activity.

# Module 6 Lecture Link:

http://ocw.mit.edu/courses/physics/8-01-physics-i-classical-mechanics-fall-1999/video-lectures/lecture-28/

# Module 6- Student's self-assessment

Please answer the following questions, which give you an indication of your standard of learning for this module.

- Could you describe pressure and flow rate?

- Could you describe buoyancy?

- Can you apply knowledge to solve problems involving fluid transfers?

- What sections of the textbook would you read for more information on the above?

- Did you enjoy your lesson?

- What aspect of the lesson was most interesting to you?

- Can you now confidently do the lesson's activities?

Now take the chapter 9 test (not included with this book)

# THERMAL PHYSICS

## Objectives

At the end of this lesson, you should be able to:

1.  Answer questions on thermal expansion
2.  Apply knowledge on heat, temperature, internal energy, and kinetic theory

## Lecture Notes

### Temperature, Internal Energy, Heat, and Specific Heat Capacity

As you know, temperature is a measure of how hot something is. Rub two sticks together and you will notice that the temperature of each increases. You did work on the sticks and their temperature increased. Doing work is transferring energy. So, you transferred energy to the sticks and their temperature increased. This means that an increase in the temperature of a system is an indication of an increase in the internal energy (a.k.a. thermal energy) of the system. (In this context, the word system is thermodynamics jargon for the generalization of the word object. Indeed an object, say an iron ball, could be a system. A system is just the subject of our investigations or considerations. A system can be as simple as a sample of one kind of gas or a chunk of one kind of metal, or it can be more complicated as in the case of a can plus some water in the can plus a thermometer in the water plus a lid on the can. For the case at hand, the system is the two sticks.) The internal energy of a system is energy associated with the motion of molecules, atoms, and the particles making up atoms relative to the center of mass of the system, and the potential energy corresponding to the positions and velocities of the aforementioned submicroscopic constituents of the system relative to each other. As usual with energy accounting, the absolute zero of energy in the case of internal energy does not matter—only changes in internal energy have any relevance. As such, you, or the publisher of a table of internal energy values (for a given substance, publishers actually list the internal energy per mass or the internal energy per mole of the substance under specified conditions rather that the internal energy of a sample of such a substance), are free to choose the zero of internal energy for a given system. In making any predictions regarding a physical process involving that system, as long as you stick with the same zero of internal energy throughout your analysis, the measurable results of your prediction or explanation will not depend on your choice of the zero of internal energy.

Another way of increasing the temperature of a pair of sticks is to bring them into contact with something hotter than the sticks are. When you do that, the temperature of the sticks automatically increases—you do not have to do any work on them. Again, the increase in the temperature of either stick indicates an increase in the internal energy of that stick. Where did that energy come from? It must have come from the hotter object. You may also notice that the hotter object's temperature decreased when you brought it into contact with the sticks. The decrease in temperature of the hotter object is an indication that the amount of internal energy in the hotter object decreased. You brought the hotter object in contact with the sticks and energy was automatically transferred from the hotter object to the sticks. The energy transfer in this case is referred to as the flow of heat. Heat is energy that is automatically transferred from a hotter object to a cooler object when you bring the two objects in contact with each other. Heat is not something that a system has but rather energy that is transferred or is being transferred. Once it gets to the system to which it is transferred we call it internal energy. The idea is to distinguish between what is being done to a system, "Work is done on the system and/or heat is caused to flow into it", with how the system changes because of what was done to it, "The internal energy of the system increases."

The fact that an increase in the temperature of an object is an indication of energy transferred to that object might suggest that anytime you transfer energy to an object its temperature increases. However, this is not the case. Try putting a hot spoon in a glass of ice water (here we consider a case for which there is enough ice so that not all of the ice melts). The spoon gets as cold as the ice water and some of the ice melts, but the temperature of the ice water remains the same (0 °C). The cooling of the spoon indicates that energy was transferred from it, and since the spoon was in contact with the ice water the energy must have been transferred to the ice water. Indeed the ice does undergo an observable change; some of it melts. The presence of more liquid water and less ice is an indication that there is more energy in the ice water. Again, there has been a transfer of energy from the spoon to the ice water. This transfer is an automatic flow of heat that takes place when the two systems are brought into contact with one another. Evidently, heat flow does not always result in a temperature increase.

Experiment shows that when a higher temperature object is in contact with a lower temperature object, heat is flowing from the higher temperature object to the lower temperature object. The flow of heat persists until the two objects are at one and the same temperature. We define the average translational kinetic energy of a molecule of a system as the sum of the translational kinetic energies of all the molecules making up the system divided by the total number of molecules. When two simple ideal gas systems, each involving a multitude of single atom molecules interacting via elastic collisions, are brought together, we find that heat flows from the system in which the average translational kinetic energy per molecule is greater to the system in which the average translational kinetic energy per molecule is lesser. This means the former system is at a higher temperature. That is to say that the higher the translational kinetic energy, on the average, of the particles making up the system, the higher the temperature. This is true for many systems.

Solids consist of atoms that are bound to neighboring atoms such that molecules tend to be held in their position, relative to the bulk of the solid, by electrostatic forces. A pair of molecules that are bound to each other has a lower amount of internal potential energy relative to the same pair of molecules when they are not bound together because we have to add energy to the bound pair at rest to yield the free pair at rest. In the case of ice water, the transfer of energy into the ice water results in the breaking of bonds between water molecules, which we see as the melting of the ice. As such, the transfer of energy into the ice water results in an increase in the internal potential energy of the system.

The two different kinds of internal energy that we have discussed are internal potential energy and internal kinetic energy. When there is a net transfer of energy into a system, and the macroscopic mechanical energy of the system doesn't change (e.g. for the case of an object near the surface of the earth, the speed of object as a whole does not increase, and the elevation of the object does not increase), the internal energy (the internal kinetic energy, the internal potential energy, or both) of the system increases. In some, but not all, cases, the increase in the internal energy is accompanied by an increase in the temperature of the system. If the temperature does not increase, then we are probably dealing with a case in which it is the internal potential energy of the system that increases.

## Heat Capacity and Specific Heat Capacity

Let us focus our attention on cases in which heat flow into a sample of matter is accompanied by an increase in the temperature of the sample. For many substances, over certain temperature ranges, the temperature change is (at least approximately) proportional to the amount of heat that flows into the substance.

$$\Delta T \infty Q,$$

the proportionality constant is 1/C, therefore:

$$Q = C\Delta T \tag{1}$$

which states that the amount of heat that must flow into a system to change the temperature of that system by $\Delta T$ is the heat capacity C times the desired temperature change $\Delta T$. Thus, the heat capacity C is the "heat-per-temperature-change." Its reciprocal is a measure of a system's temperature sensitivity to heat flow.

Let us focus our attention on the simplest kind of system, a sample of one kind of matter, such as a certain amount of water. The amount of heat that is required to change the temperature of the sample by a certain amount is directly proportional to the mass of the single substance; e.g., if you double the mass of the sample it will take twice as much heat to raise its temperature by, for instance, 1 C°. Mathematically, we can write this fact as

$$C \infty m$$

It is traditional to use a lower case $c$ for the constant of proportionality. Then

$$C = cm$$

where the constant of proportionality $c$ is the heat-capacity-per-mass of the substance in question. The heat-capacity-per-mass $c$ is referred to as the mass specific heat capacity or simply the *specific heat capacity* of the substance in question. (In this context, the adjective specific means "per amount." Because the amount can be specified in more than one way, we have the expression "mass specific" meaning "per amount of mass" and the expression "molar specific" meaning "per number of moles." Here, since we are only dealing with mass specific heat, we can omit the word "mass" without generating confusion.) The specific heat capacity c has a different value for each different kind of substance in the universe. (Okay, there might be some coincidental duplication but you get the idea.) In terms of the mass specific heat capacity, equation 1, *(Q = CAT),* for the case of a system consisting only of a sample of a single substance, can be written as

$$Q = mcAT \hspace{10cm} 2$$

The specific heat capacity *c* is a property of the kind of matter of which a substance consists. As such, the values of specific heat for various substances can be tabulated,

| Substance | Specific Heat *Capacity*<br>*J/C.Kg* |
|---|---|
| Ice (solid water) | 2090 |
| Liquid Water | 4186 |
| Water Vapor (gas) | 2000 |
| Solid Copper | 387 |
| Solid Aluminum | 902 |
| Solid Iron | 448 |

\*    The specific heat capacity of a substance varies with temperature and pressure. The values given correspond to atmospheric pressure. Use of these representative constant values for cases involving atmospheric pressure and temperature ranges between -100°C and +600°C , as applicable for the phase of the material, can be expected to yield reasonable results. However, if precision is required, or information on how reasonable your results are is needed, you should consult a thermodynamics textbook and thermodynamics tables and carry out a more sophisticated analysis.

Note how many more Joules of energy are needed to raise the temperature of 1 kg of liquid water 1 C° than are required to raise the temperature of 1 kg of a metal 1 C°.

## Temperature

Despite the fact that you are quite familiar with it, some more discussion of temperature is in order. Whenever you measure something, you are really just comparing that something with an arbitrarily-established standard. For instance, when you measure the length of a table with a meter stick, you are comparing the length of the table with the modern day equivalent of what was historically established as one ten-thousandth of the distance from the earth's north pole to the equator. In the case of temperature, a standard, now called the "degree Celsius" was established as follows: At 1 atmosphere of pressure, the temperature at which water freezes was defined to be 0 °C and the temperature at which water boils was defined to be 100 °C. Then a substance with a temperature-dependent measurable characteristic, such as the length of a column of liquid mercury, was used to interpolate and extrapolate the temperature range. (Mark the position of the end of the column of mercury on the tube containing that mercury when it is at the temperature of freezing water and again when it is at the temperature of boiling water. Divide the interval between the two marks into a hundred parts. Use the same length of each of those parts to extend the scale in both directions and call it a temperature scale.)

Note the arbitrary manner in which the zero of the Celsius scale has been established. The choice of zero is irrelevant for our purposes since equations:

$1 \rightarrow (Q = CAT)$

and

$1 \rightarrow (Q = mc \ AT)$

relate temperature *change,* rather than temperature itself, to the amount of heat flow. An absolute temperature scale has been established for the SI system of units. The zero of temperature on this scale is set at the greatest possible temperature such that it is theoretically impossible for the temperature of any system in equilibrium to be as low as the zero of the Kelvin scale. The unit of temperature on the Kelvin scale is the kelvin, abbreviated K. Note the absence of the degree symbol in the unit. The Kelvin scale is similar to the Celsius scale in that a change in temperature of, say, 1 K, is equivalent to a change in temperature of 1 C°. (Note regarding units notation: The units °C are used for a *temperature* on the Celsius scale, but the units C° are used for a *temperature change* on the Celsius scale.)

On the Kelvin scale, at a pressure of one atmosphere, water freezes at 273.15 K. So, a temperature in kelvin is related to a temperature in °C by

Temperature in K = (Temperature in °C) + 273.15

# PowerPoint Link:

Please refer to the end of the module lecture link.

# Discussion Question

Answer the following questions with references. Please remember to follow the standard APA referencing style.

For APA standards of references, please visit: http://owl.english.purdue.edu/owl/resource/560/01/

Also, respond in detail to one other post by fellow students.

**10.1** Discuss the functions of a thermo flask in terms of conduction, convection, and radiation. What improvements would you add to a modern day thermo flask?

**10.2** Describe Stefan's law with examples and discuss its applications.

# Laboratory Activity and the link (for chapters 10, 11, and 12)

Go to the link below and run the simulation:

http://phet.colorado.edu/simulations/sims.php?sim=Blackbody_Spectrum

Go to: http://phet.colorado.edu/en/contributions/view/2911

Open PhET Blackbody Spectrum.doc

Do the activity.

Now take the chapter 10 test (not included with this book)

# Module 7

## CHAPTER - 11

# Energy in Thermal Process

## Objectives

At the end of this lesson, you should be able to:

1. Answer questions on internal energy, energy transfer, phase changes
2. Apply knowledge on internal energy to answer questions on heating systems

## Lecture Notes

### Heat: Phase Changes

There is a tendency to believe that any time heat is flowing into ice, the ice is melting. NOT SO. When heat is flowing into ice, the ice will be melting only if the ice is already at the melting temperature. When heat is flowing into the ice that is below the melting temperature, the temperature of the ice is increasing.

As mentioned in the preceding chapter, there are times when you bring a hot object into contact with a cooler sample, that heat flows from the hot object to the cooler sample, but the temperature of the cooler sample does not increase, even though no heat flows out of the cooler sample (e.g. into an even colder object). This occurs when the cooler sample undergoes a phase change. For instance, if the cooler sample happens to be $H_2O$ ice or $H_2O$ ice plus liquid water, at 0°C and atmospheric pressure, when heat is flowing into the sample, the ice is melting with no increase in temperature. This will continue until all the ice is melted (assuming enough heat flows into the sample to melt all the ice). Then, after the last bit of ice melts at 0°C, if heat continues to flow into the sample, the temperature of the sample will be increasing.

In this discussion, we are treating the sample as if it had one well-defined temperature. This is an approximation. When the sample is in contact with a hotter object so that heat is flowing from the hotter object to the sample, the part of the sample in direct contact with and in the near vicinity of the hotter object will be at a higher temperature than other parts of the sample. The hotter the object, the greater the variation in the temperature of the local bit of the sample with distance from the object. We neglect this temperature variation so our discussion is only appropriate when the temperature variation is small.

Let us review the question about how it can be that heat flows into the cooler sample without causing the cooler sample to warm up. Energy flows from the hotter object to the cooler sample, but the internal kinetic energy of the cooler sample does not increase. Again, how can that be? What happens is that the energy flow into the cooler sample is accompanied by an increase in the internal potential energy of the sample. It is associated with the breaking of electrostatic bonds between molecules where the negative part of one molecule is bonded to the positive part of another. The separating of the molecules corresponds to an increase in the potential energy of the system. This is similar to a book resting on a table. It is gravitationally bound to the earth. If you lift the book and put it on a shelf that is higher than the tabletop, you have added some energy to the earth/book system, but you have increased the potential energy with no net increase in the kinetic energy. In the case of melting ice, heat flow into the sample manifests itself as an increase in the potential energy of the molecules without an increase in the kinetic energy of the molecules (which would be accompanied by a temperature increase).

The amount of heat that must flow into a single-substance solid sample that is already at its melting temperature in order to melt the whole sample depends on a property of the substance of which the sample consists, and on the mass of the substance. The relevant substance property is called the latent heat of melting. The latent heat of melting is the heat-per-mass needed to melt the substance at the melting temperature. Note that, despite the name, the latent heat is not an amount of heat but rather a ratio of heat to mass. The symbol used to represent latent heat in general is L, and we use the subscript m for melting. In terms of the latent heat of melting, the amount of heat, Q, that must flow into a sample of a single-substance solid that is at the melting temperature, in order to melt the entire sample is given by:

$$Q = mL_m$$

Note the absence of a $\Delta T$ in the expression $Q = mL_m$. There is no $\Delta T$ in the expression because there is no temperature change in the process. The whole phase change takes place at one temperature.

So far, we have talked about the case of a solid sample, at the melting temperature, which is in contact with a hotter object. Heat flows into the sample, melting it. Now consider a sample of the same substance in liquid form at the same temperature but in contact with a colder object. In this case, heat will flow from the sample to the colder object. This heat loss from the sample does not result in a decrease in the temperature. Rather, it results in a phase change of the substance of which the sample consists, from liquid to solid. This phase change is called freezing. It also goes by the name of solidification. The temperature at which freezing takes place is called the freezing temperature, but it is important to remember that the freezing temperature has the same value as the melting temperature. The heat-per-mass that must flow out of the substance to freeze it (assuming the substance to be at the freezing temperature already) is called the latent heat of fusion, or $L_f$. The latent heat of fusion for a given substance has the same value as the latent heat of melting for that substance:

$L_f = L_m$

The amount of heat that must flow out of a sample of mass m in order to convert the entire sample from liquid to solid is given by:

$Q = mL_f$

Again, there is no temperature change.

The other two phase changes we need to consider are vaporization and condensation. Vaporization is also known as boiling. It is the phase change in which liquid turns into gas. It too (as in the case of freezing and melting), occurs at a single temperature, but for a given substance, the boiling temperature is higher than the freezing temperature. The heat-per-mass that must flow into a liquid to convert it to gas is called the latent heat of vaporization $L_v$ . The heat that must flow into mass m of a liquid that is already at its boiling temperature (a.k.a. its vaporization temperature) to convert it entirely into gas is given by:

$Q = mL_v$

Condensation is the phase change in which gas turns into liquid. In order for condensation to occur, the gas must be at the condensation temperature, the same temperature as the boiling temperature (a.k.a. the vaporization temperature). Furthermore, heat must flow out of the gas, as it does when the gas is in contact with a colder object. Condensation takes place at a fixed temperature known as the condensation temperature. (The melting temperature, the freezing temperature, the boiling temperature, and the condensation temperature are also referred to as the melting point, the freezing point, the boiling point, and the condensation point, respectively.) The heat-per-mass that must be extracted from a particular kind of gas that is already at the condensation temperature, to convert that gas to liquid at the same temperature, is called the latent heat of condensation $L_c$. For a given substance, the latent heat of condensation has the same value as the latent heat of vaporization. For a sample of mass m of a gas at its condensation temperature, the amount of heat that must flow out of the sample to convert the entire sample to liquid is given by:

$Q = mL_c$

It is important to note that the actual values of the freezing temperature, the boiling temperature, the latent heat of melting, and the latent heat of vaporization are different for different substances. For water, we have:

| Phase Change | Temperature | Latent Heat |
|---|---|---|
| Melting Freezing | 0°C | MJ 0.334 kg |
| Boiling or Vaporization Condensation | 100°C | MJ 2.26 kg |

# Example 1

How much heat does it take to convert 444 grams of $H_2O$ ice at –9.0°C to steam ($H_2O$ gas) at 128.0 °C ?

## Discussion of Solution

Rather than solve this one for you, we simply explain how to solve it.

To convert the ice at $-9.0°C$ to steam at $128.0°C$, we first have to cause enough heat to flow into the ice to warm it up to the melting temperature, $0°C$. This step is a specific heat capacity problem. We use

$$Q_1 = mc_{ice}\Delta T$$

where $\Delta T$ is $[0°C - (-9.0°C)] = 9.0C°$.

Now that we have the ice at the melting temperature, we have to add enough heat to melt it. This step is a latent heat problem.

$Q_2 = mL_m$ After $Q + Q$ flows into the $H_2O$, we have liquid water at $0°C$. Next, we have to find how much heat must flow into the liquid water to warm it up to the boiling point, $100 °C$.

$$Q_3 = mc_{liquid\ water}\Delta T'$$

where $\Delta T' = (100 °C - 0°C) = 100 °C$.

After $Q_x + Q_2 + Q_3$ flows into the $H_2O$, we have liquid water at $100 °C$. Next, we have to find how much heat must flow into the liquid water at $100 °C$ to convert it to steam at $100 °C$.

$$Q_4 = mL_v$$

After $Q_x + Q_2 + Q_3 + Q_4$ flows into the $H_2O$, we have water vapor (gas) at $100 °C$. Now, all we need to do is to find out how much heat must flow into the water gas at $100 °C$ to warm it up to $128 °C$.

$$Q_5 = mc^\wedge_m \Delta T''$$

where $\Delta T''' = 128 °C - 100 °C = 28 °C$.

So, the amount of heat that must flow into the sample of solid ice at $-9.0 °C$ in order for the sample to become steam at $128 °C$ (the answer to the question) is:

$$Q_{UM} = Q + Q + Q + Q + Q$$

# PowerPoint Link:

Please refer to the end of the module lecture link.

## Discussion Question

Answer the following questions with references. Please remember to follow the standard APA referencing style.

For APA standards of references, please visit: http://owl.english.purdue.edu/owl/resource/560/01/

Also, respond in detail to one other post by fellow students.

**11.1** If one cooks rice on top of a mountain, it takes a longer time for the rice to cook than that it takes at sea level. Discuss this in terms of boiling point, phase changes etc.

**11.2** Discuss Joule-Thompson effect with relevant examples and formulae.

# Laboratory Activity and the link (for chapters 10, 11, and 12)

Go to the link below and run the simulation:

http://phet.colorado.edu/simulations/sims.php?sim=Blackbody_Spectrum

Go to: http://phet.colorado.edu/en/contributions/view/2911

Open PhET Blackbody Spectrum.doc

Do the activity.

Now take the chapter 11 test (not included with this book)

# The Laws of Thermodynamics

## Objectives

At the end of this lesson, you should be able to:

1.   Answer questions on thermodynamics

2.   Apply knowledge on thermodynamics to solve problems

## Lecture Notes

### The First Law of Thermodynamics

We use the symbol U to represent internal energy. That is the same symbol that we used to represent the mechanical potential energy of an object. Do not confuse the two different quantities with each other. In problems, questions, and discussion, the context will tell you whether the U represents internal energy or it represents mechanical potential energy.

We end this physics textbook as we began the physics part of it (Chapter 1 was a mathematics review), with a discussion of conservation of energy. Back in Chapter 2, the focus was on the conservation of mechanical energy; here we focus our attention on thermal energy.

In the case of a deformable system, it is possible to do some net work on the system without causing its mechanical kinetic energy $\frac{1}{2}mv^2 + \frac{1}{2}Iw^2$ to change (where m is the mass of the system, $v$ is: the speed of the center of mass of the system, I is the moment of inertia of the system, and w is the magnitude of the angular velocity of the system). Examples of such work would be: the bending of a coat hanger, the stretching of a rubber band, the squeezing of a lump of clay, the compression of a gas, and the stirring of a fluid.

When you do work on something, you transfer energy to that something. For instance, consider a case in which you push on a cart that is initially at rest. Within your body, you convert chemical potential energy into mechanical energy, which, by pushing the cart, you give to the cart. After you have been pushing on it for a while, the cart is moving, meaning that it has some kinetic energy. So, in the end, the cart has some kinetic energy that was originally chemical potential energy stored in you. Energy has been transferred from you to the cart.

In the case of the cart, what happens to the energy that you transfer to the cart is clear. Yet, what about the case of a deformable system whose center of mass stays put? When you do work on such a system, you transfer energy to that system. So what happens to the energy? Experimentally, we find that the energy becomes part of the internal energy of the system. The internal energy of the system increases by an amount that is equal to the work done on the system.

This increase in the internal energy can be an increase in the internal potential energy, an increase in the internal kinetic energy, or both. An increase in the internal kinetic energy would manifest itself as an increase in temperature.

Doing work on a system represents the second way, which we have considered, of causing an increase in the internal energy of the system. The other way was for heat to flow into the system. The fact that doing work on a system and/or having heat flow into that system will increase the internal energy of that system, is represented, in equation form, by:

$$\Delta U = Q + W_{IN}$$

which we copy here for your convenience:

$$\Delta U = Q + W_{IN}$$

In this equation, $\Delta U$ is the change in the internal energy of the system, $Q$ is the amount of heat that flows into the system, and WIN is the amount of work that is done on the system. This equation is referred to as the First Law of Thermodynamics. Chemists typically write it without the subscript IN on the symbol W representing the work done on the system. (The subscript IN is there to remind us that the $W_{IN}$ represents a transfer of energy into the system. In the chemistry convention, it is understood that W represents the work done on the system—no subscript is necessary.)

Historically, physicists and engineers have studied and developed thermodynamics with the goal of building a better heat engine, a device, such as a steam engine, designed to produce work from heat. That is, a device for which heat goes in and work comes out. It is probably for this reason that physicists and engineers almost always write the first law as:

$$\Delta U = Q - W$$

where the symbol W represents the amount of work done by the system on the external world. (This is just the opposite of the chemistry convention.) Because this is a physics course, this ($\Delta U = Q - W$) is the form in which the first law appears on your formula sheet. It is suggested that the first law be as explicit as possible by writing it as $\Delta U = Q_{IN} - W_{OUT}$ or, better yet:

$$\Delta U = Q_{IN} + W_{IN}$$

In this form, the equation is saying that you can increase the internal energy of a system by causing heat to flow into that system and/or by doing work on that system. Note that any one of the quantities in the equation can be negative. A negative value of $Q_{IN}$ means that heat actually flows out of the system. A negative value of W$_{IN}$ means that work is actually done by the system on the surroundings. Finally, a negative value of $\Delta U$ means that the internal energy of the system decreases.

Again, the real key is to use subscripts and common sense. Write the First Law of Thermodynamics in a manner consistent with the facts that heat or work into a system will increase the internal energy of the system, and heat or work out of the system will decrease the internal energy of the system.

# Second Law of Thermodynamics

1.  Heat Engines

    A *heat engine* is a cyclic device that takes heat Q$_H$ in from a *hot reservoir*, converts *some* of it to work W, and rejects the rest of it to Q$_c$ a *cold reservoir*, so that at the end of a cycle it is in the same state (and has the same internal energy). The net work done per cycle is (recall) the area inside the PV curve.

    The *efficiency* of a heat engine is defined as,

    $$\varepsilon = \frac{W}{Q_H} = \frac{Q_H - Q_C}{Q_H} = 1 - \frac{Q_C}{Q_H}$$

2.  Refrigerators (and Heat Pumps)

    A *refrigerator* is basically a cyclic heat engine that operates backwards. In a cycle it takes heat in from a cold reservoir, does work W, on it, and rejects a heat Q$_H$ to a hot reservoir. Its net effect is to make the cold reservoir colder (refrigeration) by removing heat from inside it to the warmer warm reservoir (warming it further, e.g. as a heat pump).

    The *coefficient of performance* of a refrigerator is defined to be,

    $$COP = \frac{Q_C}{W}$$

3.  Carnot Engine

    The *Carnot Cycle* is the archetypical reversible cycle, and a Carnot Cycle-based heat engine is one that does not dissipate any energy internally and uses only reversible steps. *Carnot's Theorem* states that no real heat engine operating between a hot reservoir at temperature T$_H$ and a cold reservoir at temperature T$_c$ can be more efficient than a Carnot engine operating between those two reservoirs.

A Carnot Cycle consists of four steps:

1.  Isothermal expansion (in contact with the heat reservoir)

2.  Adiabatic expansion (after the heat reservoir is removed)

3.  Isothermal compression (in contact with the cold reservoir)

4.  Adiabatic compression (after the cold reservoir is removed)

The efficiency of a Carnot Engine is:

$$\varepsilon_{\text{Carnot}} = 1 - \frac{T_C}{T_H}$$

## Entropy

Entropy, S, is a measure of disorder. The change in entropy of a system can be evaluated by integrating:

$$dS = \frac{dQ}{T}$$

A reversible process is one where the entropy of the system does not change. An irreversible process increases the entropy of the system and its surroundings.

## Entropy Statement of the Second Law of Thermodynamics

The entropy of the Universe never decreases. It either increases (for irreversible processes) or remains the same (for reversible processes).

# PowerPoint Link:

Please refer to the end of the module lecture link.

# Discussion Question

Answer the following questions with references. Please remember to follow the standard APA referencing style.

For APA standards of references, please visit: http://owl.english.purdue.edu/owl/resource/560/01/

Also, respond in detail to one other post by fellow students.

12.1 Visit the following website: http://www.wired.com/gadgetlab/2008/08/vertical-windsp/

Discuss the claims in terms of Carnot engine and other thermodynamic principles.

12.2 Derive relevant equations to find the work done by an ideal gas during reversible and irreversible expansions.

# Laboratory Activity and the link (for chapters 10, 11, and 12)

Go to the link below and run the simulation:

http://phet.colorado.edu/simulations/sims.php?sim=Blackbody_Spectrum

Go to: http://phet.colorado.edu/en/contributions/view/2911

Open PhET Blackbody Spectrum.doc

Do the activity.

# Module 7 Lecture Links:

http://ocw.mit.edu/courses/physics/8-01-physics-i-classical-mechanics-fall-1999/video-lectures/lecture-32/

http://ocw.mit.edu/courses/physics/8-01-physics-i-classical-mechanics-fall-1999/video-lectures/lecture-33/

# Module 7 – Student's self-assessment

**Please answer the** following **questions, which give you an indication of your standard of learning for this module.**

- Could you describe heat and temperature?

- Could you describe the three laws of thermodynamics?

- Can you apply knowledge to solve problems involving energy transfers?

- What sections of the textbook would you read for more information on the above?

- Did you enjoy your lesson?

- What aspect of the lesson was most interesting to you?

- Can you now confidently do the lesson's activities?

Now take the chapter 12 test (not included with this book)

## CHAPTER - 13

# Vibrations and Waves

## Objectives

At the end of this lesson, you should be able to:

1. Answers questions on vibrations and waves
2. Apply waves and vibrations equations to solve problems

## Lecture Notes

### Oscillations: Introduction, Mass on a Spring

If a simple harmonic oscillation problem does not involve the time, you should probably be using conservation of energy to solve it. A common "tactical error" in problems involving oscillations is to manipulate the equations giving the position and velocity as a function of time, $x = x_{max} \, \text{Cos}(2\pi f \, t)$ and $v = -V_{max} \, \text{Sin}(2\pi f \, t)$ rather than applying the principle of conservation of energy.

When something goes back and forth we say it vibrates or oscillates. In many cases, oscillations involve an object whose position as a function of time is well characterized by the sine or cosine function of the product of a constant and elapsed time. Such motion is referred to as sinusoidal oscillation. It is also referred to as simple harmonic motion.

### The Cosine Function

By now, you have had a great deal of experience with the cosine function of an angle as the ratio of the adjacent to the hypotenuse of a right triangle. This definition covers angles from 0 radians to $\dfrac{\pi}{2}$ radians (0° to 90°). In applying the cosine function to simple harmonic motion, we use the extended definition which covers all angles. The extended definition of the cosine of the angle $\theta$ is that the cosine of an angle is the $x$ component of a unit vector, the tail of which is on the origin of an $x$-$y$ coordinate system; a unit vector that originally pointed in the $+x$ direction but has since been rotated counterclockwise from our viewpoint, through the angle $\theta$, about the origin.

Here we show that the extended definition is consistent with the "adjacent over hypotenuse" definition, for angles between 0 radians and $\dfrac{\pi}{2}$ radians.

For such angles, we have:

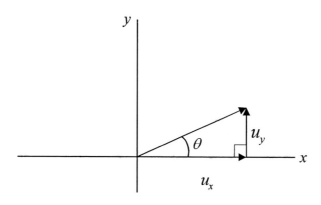

**Figure 13.1**

in which, $u$, being the magnitude of a unit vector, is of course equal to 1, the pure number 1 with no units. Now, according to the ordinary definition of the cosine of $\theta$ as the adjacent over the hypotenuse:

$$\cos\theta = \frac{u_x}{u}$$

Solving this for $u_x$ we see that

$$u_x = u\cos\theta$$

Recalling that $u = 1$, this means that

$$u_x = \cos\theta$$

Recalling that our extended definition of $\cos\theta$, that it is the $x$ component of the unit vector $\hat{u}$ when $\hat{u}$ makes an angle $\theta$ with the $x$-axis, this last equation is just saying that, for the case at hand ($\theta$ between 0 and $\dfrac{\pi}{2}$ radians) our extended definition of $\cos\theta$ is equivalent to our ordinary definition.

At angles between $\dfrac{\pi}{2}$ and $\dfrac{3\pi}{2}$ radians (90° and 270°) we see that $u_x$ takes on negative values (when the $x$ component *vector* is pointing in the negative $x$ direction, the $x$ component *value* is, by definition, negative).   According to our extended definition, cos $\theta$ takes on negative values at such angles as well.

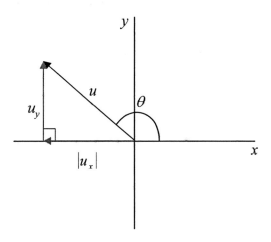

**Figure 13.2**

With our extended definition, valid for any angle $\theta$ , a graph of the cos $\theta$ versus $\theta$ appears as:

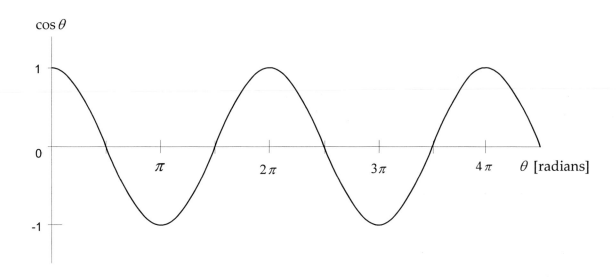

**Figure 13.3**

## Some Calculus Relations Involving the Cosine

The derivative of the cosine of $\theta$, with respect to $\theta$ :

$$\frac{d}{d\theta}\cos\theta = -\sin\theta$$

The derivative of the sine of $\theta$ , with respect to $\theta$ :

$$\frac{d}{d\theta}\sin\theta = \cos\theta$$

## Some Jargon Involving The Sine And Cosine Functions

When you express, define, or evaluate the function of something, that something is called the argument of the function. For instance, suppose the function is the square root function and the expression in question is $\sqrt{3x}$. The expression is the square root of 3x, so, in that expression, 3x is the argument of the square root function. Now when you take the cosine of something, that something is called the argument of the cosine, but in the case of the sine and cosine functions, we give it another name as well, namely, the phase. So, when you write $\cos\theta$, the variable $\theta$ is the argument of the cosine function, but it is also referred to as the *phase* of the cosine function. In order for an expression involving the cosine function to be at all meaningful, the phase of the cosine must have units of angle (for instance, radians or degrees).

## A Block Attached to the End of a Spring

Consider a block of mass $m$ on a frictionless horizontal surface. The block is attached, by means of an ideal massless horizontal spring having force constant $k$, to a wall. A person has pulled the block out, directly away from the wall, and released it from rest. The block oscillates back and forth (toward and away from the wall), on the end of the spring. We would like to find equations that give the block's position, velocity, and acceleration as functions of time. We start by applying Newton's second law to the block. Before drawing the free body diagram we draw a sketch to help identify our one-dimensional coordinate system. We will call the horizontal position of the point at which the spring is attached, the position $x$ of the block. The origin of our coordinate system will be the position at which the spring is neither stretched nor compressed. When the position $x$ is positive, the spring is stretched and exerts a force, on the block, in the $-x$ direction. When the position of $x$ is negative, the spring is compressed and exerts a force, on the block, in the $+x$ direction.

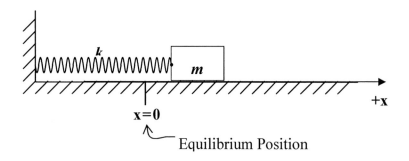

**Figure 13.4**

Now we draw the free body diagram of the block:

and apply Newton's second law:

$$a_\rightarrow = \frac{1}{m}\sum F_\rightarrow$$

$$a = \frac{1}{m}(-kx)$$

$$a = -\frac{k}{m}x$$

This equation, relating the acceleration of the block to its position $x$, can be considered to be an equation relating the position of the block to time if we substitute for $a$ using:

$$a = \frac{dv}{dt}$$

and

$$v = \frac{dx}{dt}$$

so

$$a = \frac{d}{dt}\frac{dx}{dt}$$

which is usually written

$$a = \frac{d^2x}{dt^2}$$

1

Substituting this expression for $a$ into $a = -\frac{k}{m}x$ (the result we derived from Newton's second law above) yields,

$$\frac{d^2x}{dt^2} = -\frac{k}{m}x$$

2

We know in advance that the position of the block depends on time. That is to say, $x$ is a function of time. This equation, **Error! Reference source not found.** tells us that if you take the second derivative of $x$ with respect to time you get $x$ itself, times a negative constant ($-k/m$).

We can find an expression for x in terms of t that solves equation by the method of "guess and check." Grossly, we are looking for a function whose second derivative is essentially the negative of itself. Two functions meet this criterion, the sine and the cosine. Either will work. We arbitrarily choose to use the cosine function. We include some constants in our trial solution (our guess) to be determined during the "check" part of our procedure. Here is our trial solution:

$$x = x_{max} \cos\left(\frac{2\pi \text{ rad}}{T}t\right)$$

Here is how we have arrived at this trial solution: Having established that x depends on the cosine of a multiple of the time variable, we let the units be our guide. We need the time t to be part of the argument of the cosine, but we cannot take the cosine of something unless that something has units of angle. The constant $\frac{2\pi \text{ rad}}{T}$, with the constant T having units of time (we will use seconds), makes it so that the argument of the cosine has units of radians. It is, however, more than just the units that motivate us to choose the ratio $\frac{2\pi \text{ rad}}{T}$ as the constant. To make the argument of the cosine have units of radians, all we need is a constant with units of radians per second. So why write it as $\frac{2\pi \text{ rad}}{T}$? Here's the explanation: The block goes back and forth. That is, it repeats its motion over and over again as time goes by. Starting with the block at its maximum distance from the wall, the block moves toward the wall, reaches its closest point of approach to the wall and then comes back out to its maximum distance from the wall. At that point, I i's right back where it started from. We define the constant value of time T to be the amount of time that it takes for one iteration of the motion.

Now consider the cosine function. We chose it because its second derivative is the negative of itself, but it is looking better and better as a function that gives the position of the block as a function of time because it too repeats itself as its phase (the argument of the cosine) continually increases. Suppose the phase starts out as 0 at time 0. The cosine of 0 radians is 1, the biggest the cosine ever gets. We can make this correspond to the block being at its maximum distance from the wall. As the phase increases, the cosine gets smaller, then goes negative, eventually reaching the value -1 when the phase is π radians. This could correspond to the block being closest to the wall. Then, as the phase continues to increase, the cosine increases until, when the phase is 2π, the cosine is back up to 1 corresponding to the block being right back where it started from. From here, as the phase of the cosine continues to increase from 2π to 4π, the cosine again takes on all the values that it took on from 0 to 2π. The same thing happens again as the phase increases from 4π to 6π, from 8π to 10π, etc.

Getting back to that constant $\frac{2\pi \text{ rad}}{T}$ that we "guessed" should be in the phase of the cosine in our trial solution for x(t):

$$x = x_{max} \cos\left(\frac{2\pi \text{ rad}}{T}t\right)$$

With T being defined as the time it takes for the block to go back and forth once, look what happens to the phase of the cosine as the stopwatch reading continually increases. Starting from 0, as t increases from 0 to T, the phase of the cosine, $\frac{2\pi \text{ rad}}{T}t$, increases from 0 to 2π radians. So, just as the block, from time 0 to time $T$, goes though one cycle of its motion, the cosine, from time 0 to time $T$, goes through one cycle of its pattern. As the stopwatch reading increases from $T$ to $2T$, the phase of the cosine increases from $2\pi$ rad to $4\pi$ rad. The block undergoes the second cycle of its motion and the cosine function used to determine the position of the block goes through the second cycle of its pattern. The idea holds true for any time $t$ —as the stopwatch reading continues to increase, the cosine function keeps repeating its cycle in exact synchronization with the block, as it must if its value is to accurately represent the position of the block as a function of time. Again, it is no coincidence. We chose the constant $\dfrac{2\pi \text{ rad}}{T}$ in the phase of the cosine so that things would work out this way.

A few words on jargon are in order before we move on. The time $T$ that it takes for the block to complete one full cycle of its motion is referred to as the *period* of the oscillations of the block.

Now how about that other constant, the "$x_{max}$" in our educated guess $x = x_{max} \cos\left(\dfrac{2\pi \text{ rad}}{T}t\right)$ ? Again, the units were our guide. When you take the cosine of an angle, you get a pure number, a value with no units. So, we need the $x_{max}$ there to give our function units of distance (we'll use meters). We can further relate $x_{max}$ to the motion of the block. The biggest the cosine of the phase can ever get is 1, thus, the biggest $x_{max}$ times the cosine of the phase can ever get is $x_{max}$. So, in the expression $x = x_{max} \cos\left(\dfrac{2\pi \text{ rad}}{T}t\right)$, with $x$ being the position of the block at any time $t$, $x_{max}$ must be the maximum position of the block, the position of the block, relative to its equilibrium position, when it is as far from the wall as it ever gets.

Okay, we've given a lot of reasons why $x = x_{max}\cos\left(\dfrac{2\pi \text{ rad}}{T}t\right)$ should well describe the motion of the block, but unless it is consistent with Newton's 2nd Law, that is, unless it satisfies equation**Error! Reference source not found.**:

$$\frac{d^2x}{dt^2} = -\frac{k}{m}x$$

which we derived from Newton's second law, it is no good. So, let's plug it into equation **Error! Reference source not found.** and see if it works. First, let's take the second derivative $\dfrac{d^2x}{dt^2}$ of our trial solution with respect to $t$ (so we can plug it and $x$ itself directly into equation):

Given

$$x = x_{max}\cos\left(\frac{2\pi \text{ rad}}{T}t\right),$$

the first derivative is

$$\frac{dx}{dt} = x_{max}\left[-\sin\left(\frac{2\pi \text{ rad}}{T}t\right)\right]\frac{2\pi \text{ rad}}{T}$$

$$\frac{dx}{dt} = -\frac{2\pi \text{ rad}}{T}x_{max}\sin\left(\frac{2\pi \text{ rad}}{T}t\right)$$

The second derivative is then

$$\frac{d^2x}{dt^2} = -\frac{2\pi \text{ rad}}{T}x_{max}\cos\left(\frac{2\pi \text{ rad}}{T}t\right)\frac{2\pi \text{ rad}}{T}$$

$$\frac{d^2x}{dt^2} = -\left(\frac{2\pi \text{ rad}}{T}\right)^2 x_{max}\cos\left(\frac{2\pi \text{ rad}}{T}t\right)$$

Now we are ready to substitute this and $x$ itself, $x = x_{max}\cos\left(\frac{2\pi \text{ rad}}{T}t\right)$, into the differential equation

$$\frac{d^2x}{dt^2} = -\frac{k}{m}x$$ (equation stemming from Newton's second law of motion.) The substitution yields:

$$-\left(\frac{2\pi \text{ rad}}{T}\right)^2 x_{max}\cos\left(\frac{2\pi \text{ rad}}{T}t\right) = -\frac{k}{m}x_{max}\cos\left(\frac{2\pi \text{ rad}}{T}t\right)$$

which we copy here for your convenience.

$$-\left(\frac{2\pi \text{ rad}}{T}\right)^2 x_{max}\cos\left(\frac{2\pi \text{ rad}}{T}t\right) = -\frac{k}{m}x_{max}\cos\left(\frac{2\pi \text{ rad}}{T}t\right)$$

The two sides are the same, by inspection, except that where $\left(\frac{2\pi \text{ rad}}{T}\right)^2$ appears on the left, we have $\frac{k}{m}$ on the right. Thus, substituting our guess, $x = x_{max}\cos\left(\frac{2\pi \text{ rad}}{T}t\right)$, into the differential equation that we are trying to solve, $\frac{d^2x}{dt^2} = -\frac{k}{m}x$ leads to an identity if and only if $\left(\frac{2\pi \text{ rad}}{T}\right)^2 = \frac{k}{m}$. This means that the period $T$ is determined by the characteristics of the spring and the block, more specifically by the force constant (the "stiffness factor") $k$ of the spring, and the mass (the inertia) of the block. Let us solve for $T$ in terms of these quantities. From $\left(\frac{2\pi \text{ rad}}{T}\right)^2 = \frac{k}{m}$ we find:

$$\frac{2\pi \text{ rad}}{T} = \sqrt{\frac{k}{m}}$$

$$T = 2\pi \text{ rad}\sqrt{\frac{m}{k}}$$

$$T = 2\pi\sqrt{\frac{m}{k}}$$

where we have taken advantage of the fact that a radian is, by definition, 1 m/m by simply deleting the "rad" from our result.

The presence of the $m$ in the numerator means that the greater the mass, the longer the period. That makes sense: we would expect the block to be more "sluggish" when it has more mass. On the other hand, the presence of the $k$ in the *denominator* means that the stiffer the spring, the shorter the period. This makes sense too in that we would expect a stiff spring to result in quicker oscillations. Note the absence of $x_{max}$ in the result for the period $T$. Some would expect that the bigger the oscillations, the longer it would take the block to complete each oscillation, but the absence of $x_{max}$ in our result for $T$ shows that this is not the case. The period $T$ does not depend on the size of the oscillations.

So, our end result is that a block of mass $m$, on a frictionless horizontal surface, a block that is attached to a wall by an ideal massless horizontal spring, and released, at time $t = 0$, from rest, from a position $x = x_{max}$, a distance $x_{max}$ from its equilibrium position, will oscillate about the equilibrium position with a period $T = 2\pi\sqrt{\dfrac{m}{k}}$. Furthermore, the block's position as a function of time will be given by

$$x = x_{max} \cos\left(\frac{2\pi \text{ rad}}{T} t\right)$$

2

From this expression for $x(t)$ we can derive an expression for the velocity $dx(t)$ as follows:

$$v = \frac{dx}{dt}$$

$$v = \frac{d}{dt}\left[x_{max} \cos\left(\frac{2\pi \text{ rad}}{T} t\right)\right]$$

$$v = x_{max}\left[-\sin\left(\frac{2\pi \text{ rad}}{T} t\right)\right]\frac{2\pi \text{ rad}}{T}$$

$$v = -x_{max}\frac{2\pi \text{ rad}}{T}\sin\left(\frac{2\pi \text{ rad}}{T} t\right)$$

3

And from this expression for $dx(t)$ we can get the acceleration $a(t)$ as follows:

$$a = \frac{dv}{dt}$$

$$a = \frac{d}{dt}\left[-x_{max}\frac{2\pi \text{ rad}}{T}\sin\left(\frac{2\pi \text{ rad}}{T} t\right)\right]$$

$$a = -x_{max}\frac{2\pi \text{ rad}}{T}\left[\cos\left(\frac{2\pi \text{ rad}}{T} t\right)\right]\frac{2\pi \text{ rad}}{T}$$

$$a = -x_{max}\left(\frac{2\pi \text{ rad}}{T}\right)^2 \cos\left(\frac{2\pi \text{ rad}}{T} t\right)$$

4

Note that this latter result is consistent with the relation $a = -\dfrac{k}{m}x$ between $a$ and $x$ that we derived from Newton's second law near the start of this chapter.    Recognizing that the $x_{max}\cos\left(\dfrac{2\pi\text{ rad}}{T}t\right)$ is $x$ and that the $\left(\dfrac{2\pi\text{ rad}}{T}\right)^2$ is $\dfrac{k}{m}$, it is clear that equation 4 is the same thing as

$$a = -\frac{k}{m}x \qquad\qquad 5$$

## Frequency

The period $T$ has been defined to be the time that it takes for one complete oscillation.  In SI units we can think of it as the number of seconds per oscillation.  The reciprocal of $T$ is thus the number of oscillations per second.  This is the rate at which oscillations occur.  We give it a name, frequency, and a symbol, $f$.

$$f = \frac{1}{T} \qquad\qquad 6$$

The units work out to be $\dfrac{1}{\text{s}}$ which we can think of as $\dfrac{\text{oscillations}}{\text{s}}$ as the oscillation, much like the radian is a marker rather than a true unit.  A special name has been assigned to the SI unit of frequency, $1\dfrac{\text{oscillation}}{\text{s}}$ is defined to be 1 hertz, abbreviated 1 Hz.  You can think of 1 Hz as either $1\dfrac{\text{oscillation}}{\text{s}}$ or simply $1\dfrac{1}{\text{s}}$.

In terms of frequency, rather than period, we can use $f = \dfrac{1}{T}$ to express all our previous results in terms of $f$ rather than $t$.

$$f = \frac{1}{2\pi}\sqrt{\frac{k}{m}}$$

$$x = x_{max}\cos\left(2\pi\text{ rad}f\,t\right)$$

$$v = -2\pi f\,x_{max}\sin\left(2\pi\text{ rad}f\,t\right)$$

$$a = -(2\pi f)^2\,x_{max}\cos\left(2\pi\text{ rad}f\,t\right)$$

By inspection of the expressions for the velocity and acceleration above we see that the greatest possible value for the velocity is $2\pi f x_{max}$ and the greatest possible value for the acceleration is $(2\pi f)^2 x_{max}$. Defining

$$V_{max} = x_{max}(2\pi f) \qquad\qquad 7$$

and

$$a_{max} = x_{max}(2\pi f)^2 \qquad\qquad 8$$

and, omitting the units "rad" from the phase (thus burdening the user with remembering that the units of the phase are radians while making the expressions a bit more concise) we have:

$$x = x_{max}\cos(2\pi f t) \qquad\qquad 9$$

$$v = -v_{max}\sin(2\pi f t) \qquad\qquad 10$$

$$a = -a_{max}\cos(2\pi f t) \qquad\qquad 11$$

# The Simple Harmonic Equation

When the motion of an object is sinusoidal as in $x = x_{max}\cos(2\pi f t)$, we refer to the motion as simple harmonic motion. In the case of a block on a spring we found that

$$a = -|\text{constant}|x \qquad\qquad 12$$

where the |constant| was $\dfrac{k}{m}$ and was shown to be equal to $(2\pi f)^2$. Written as

$$\frac{d^2 x}{dt^2} = -(2\pi f)^2 x \qquad\qquad 13$$

the equation is a completely general equation, not specific to a block on a spring. Indeed, any time you find that, for any system, the second derivative of the position variable, with respect to time, is equal to a negative constant times the position variable itself, you are dealing with a case of simple harmonic motion, and you can equate the absolute value of the constant to $(2\pi f)^2$.

# Oscillations: The Simple Pendulum, Energy in Simple Harmonic Motion

*Starting with the pendulum bob at its highest position on one side, the period of oscillations is the time it takes for the bob to swing all the way to its highest position on the other side and back again. Do not forget that part about "and back again."*

By definition, a simple pendulum consists of a particle of mass $m$ suspended by a massless, un-stretchable string of length $L$ in a region of space in which there is a uniform constant gravitational field, e.g. near the surface of the earth. The suspended particle is called the pendulum bob. Here we discuss the motion of the bob. While the results to be revealed here are most precise for the case of a point particle, they are good as long as the length of the pendulum (from the fixed top end of the string to the center of mass of the bob) is large compared to a characteristic dimension (such as the diameter if the bob is a sphere or the edge length if it is a cube) of the bob. (Using a pendulum bob whose diameter is 10% of the length of the pendulum (as opposed to a point particle) introduces a 0.05% error. You have to make the diameter of the bob 45% of the pendulum length to get the error up to 1%.)

If you pull the pendulum bob to one side and release it, you find that it swings back and forth. It oscillates. At this point, you do not know whether or not the bob undergoes simple harmonic motion, but you certainly know that it oscillates. To find out if it undergoes simple harmonic motion, all we have to do is to determine whether its acceleration is a negative constant times its position. Because the bob moves on an arc rather than a line, it is easier to analyze the motion using angular variables.

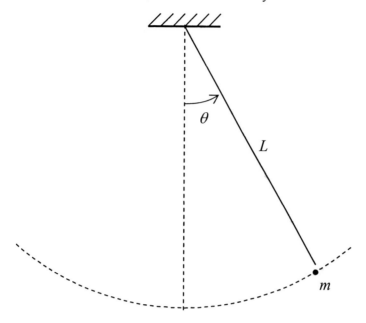

**Figure 13.5**

The bob moves on the lower part of a vertical circle that is centered at the fixed upper end of the string. We will position ourselves such that we are viewing the circle, face on, and adopt a coordinate system, based on our point of view, which has the reference direction straight downward, and for which positive angles are measured counterclockwise from the reference direction. Referring to the diagram above, we now draw a pseudo free-body diagram (the kind we use when dealing with torque) for the string-plus-bob system.

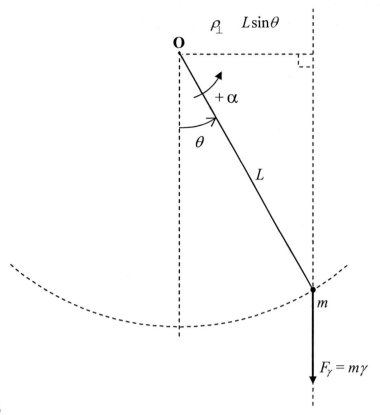

**Figure 13.6**

We consider the counterclockwise direction to be the positive direction for all the rotational motion variables. Applying Newton's second law for rotational motion, yields:

$$\alpha = \frac{\Sigma\tau}{I}$$

$$\alpha = -\frac{mgLsin\theta}{I}$$

Next we implement the small angle approximation. Doing so means our result is approximate and that the smaller the maximum angle achieved during the oscillations, the better the approximation. According to the small angle approximation, with it understood that $\theta$ must be in radians, $\sin\theta \approx \theta$. Substituting this into our expression for $\alpha$, we obtain:

$$\alpha = -\frac{mgL\theta}{I}$$

Here comes the part where we treat the bob as a point particle. The moment of inertia of a point particle, with respect to an axis that is a distance $L$ away, is given by $I = mL^2$. Substituting this into our expression for $\alpha$ we arrive at:

$$\alpha = -\frac{mgL}{mL^2}\theta$$

Something profound occurs in our simplification of this equation. The masses cancel out. The mass that determines the driving force behind the motion of the pendulum (the gravitational force $F_g = mg$) in the numerator, is exactly canceled by the inertial mass of the bob in the denominator. The motion of the bob does not depend on the mass of the bob! Simplifying the expression for $\alpha$ yields:

$$\alpha = -\frac{g}{L}\,\theta$$

Recalling that $\alpha \equiv \frac{d^2\theta}{dt^2}$, we have:

$$\frac{d^2\theta}{dt^2} = -\frac{g}{L}\,\theta$$

This is the simple harmonic motion equation.

Solving this for frequency (*f*) we find that the frequency of oscillations of a simple pendulum is given by

$$f = \frac{1}{2\pi}\sqrt{\frac{g}{L}}$$

# Energy Considerations in Simple Harmonic Motion

Let us return our attention to the block on a spring. A person pulls the block out away from the wall a distance xmax from the equilibrium position, and releases the block from rest. At that instant, before the block picks up any speed at all, (but when the person is no longer affecting the motion of the block) the block has a certain amount of energy E. Since we are dealing with an ideal system (no friction, no air resistance) the system has that same amount of energy from then on. In general, while the block is oscillating, the energy, E = K + U is partly kinetic energy $K = \frac{1}{2}mv^2$ and partly spring potential energy $U = \frac{1}{2}kx^2$. The amount of each varies, but the total remains the same. At time 0, the K in E = K + U is zero since the velocity of the block is zero. So, at time 0:

$$E = U$$

$$E = \tfrac{1}{2}kx_{max}^2$$

An endpoint in the motion of the block is a particularly easy position at which to calculate the total energy since all of it is potential energy.

As the spring contracts, pulling the block toward the wall, the speed of the block increases so, the kinetic energy increases while the potential energy $U = \frac{1}{2}kx^2$ decreases because the spring becomes less and less stretched. On its way toward the equilibrium position, the system has both kinetic and potential energy.

# Waves: Characteristics, Types, Energy

Consider a long taut horizontal string of great length. Suppose one end is in the hand of a person and the other is fixed to an immobile object. Now suppose that the person moves her hand up and down. The person causes her hand, and her end of the string, to oscillate up and down. To discuss what happens, we, in our mind, consider the string to consist of a large number of very short string segments. It is important to keep in mind that the force of tension of a string segment exerted on any object, including another segment of the string, is directed away from the object *along the string segment* that is exerting the force. (The following discussion and diagrams are intentionally oversimplified. The discussion *does* correctly give the gross idea of how oscillations at one end of a taut string can cause a pattern to move along the length of the string despite the fact that the individual bits of string are essentially doing nothing more than moving up and down.

The person is holding one end of the first segment. She first moves her hand upward.

**Figure 13.7**

This tilts the first segment so that the force of tension that it is exerting on the second segment has an upward component.

This, in turn, tilts the second segment so that its force of tension on the third segment now has an upward component. The process continues with the third segment, the forth segment, etc.

After reaching the top of the oscillation, the person starts moving her hand downward. She moves the left end of the first segment downward, but by this time, the first four segments have an upward velocity. Due to their inertia, they continue to move upward. The downward pull of the first segment on the left end of the second segment causes it to slow down, come to rest,

and eventually start moving downward. Inertia plays a huge role in wave propagation. "To propagate" means "to go" or "to travel." Waves propagate through a medium.

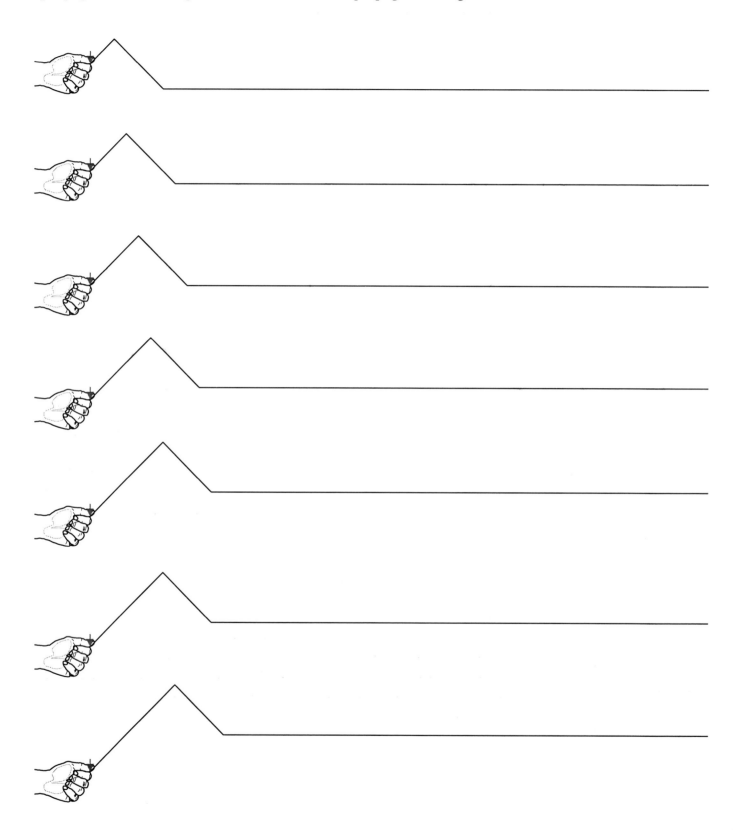

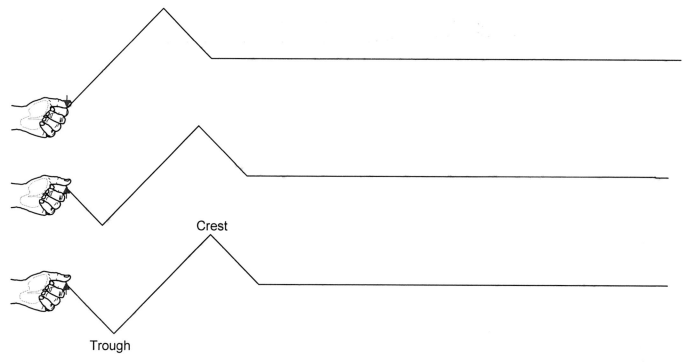

Crest

Trough

Each very short segment of the string undergoes oscillatory motion like that of the hand, but for any given section, the motion is delayed relative to the motion of the neighboring segment that is closer to the hand. The net effect of all these string segments oscillating up and down, each with the same frequency but slightly out of synchronization with its nearest neighbor, is to create a disturbance in the string. Without the disturbance, the string would just remain on the original horizontal line. The disturbance moves along the length of the string, away from the hand. The disturbance is called a wave. An observer, looking at the string from the side sees crests and troughs of the disturbance, moving along the length of the string, away from the hand. Despite appearances, no material is moving along the length of the string, just a disturbance. The illusion that actual material is moving along the string can be explained by the timing with which the individual segments move up and down, each about its own equilibrium position, the position it was in before the person started making waves.

## Wave Characteristics

In our pictorial model above, we depicted a hand that was oscillating but not undergoing simple harmonic motion. If the oscillations that are causing the wave do conform to simple harmonic motion, then each string segment making up the string will experience simple harmonic motion (up and down). When individual segments making up the string are each undergoing simple harmonic motion, the wave pattern is said to vary sinusoidally in both time and space. We can tell that it varies sinusoidally in space because a graph of the displacement $y$, the distance that a given point on the string is above its equilibrium position, versus $x$, how far from the end of the string the point on the string is; for all points on the string; is sinusoidal.

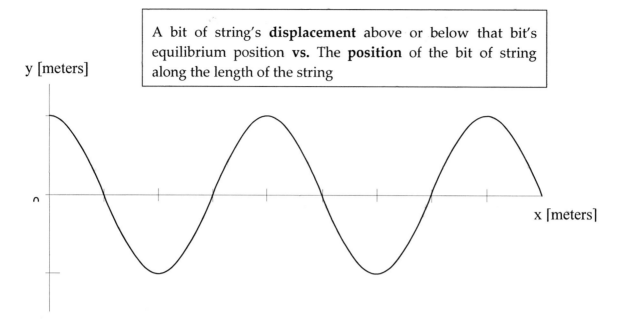

A bit of string's **displacement** above or below that bit's equilibrium position **vs.** The **position** of the bit of string along the length of the string

**Figure 13.8**

We say that the wave varies sinusoidally with time because, for any point along the length of the string, a graph of the displacement of that point from its equilibrium position versus time is sinusoidal:

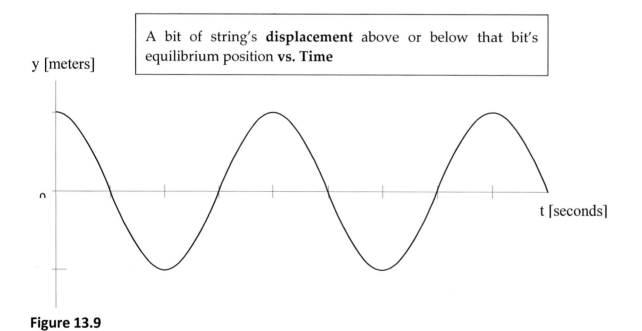

A bit of string's **displacement** above or below that bit's equilibrium position **vs. Time**

**Figure 13.9**

There are a number of ways of characterizing the wave-on-a-string system. You could probably come up with a rather complete list yourself: the rate at which the oscillations are occurring, how long it takes for a given tiny segment of the string to complete one oscillation, how big the oscillations are, the smallest length of the unique pattern which repeats itself in space, and the speed at which the wave pattern travels along the length of the string. Physicists have, of course, given names to the various quantities, in accordance with that important lowest level of scientific activity—naming and categorizing the various characteristics of that aspect of the natural world which is under study. Here are the names:

# Amplitude

Any particle of a string with waves traveling through it undergoes oscillations. Such a particle goes away from its equilibrium position until it reaches its maximum displacement from its equilibrium position. Then it heads back toward its equilibrium position and then passes right through the equilibrium position on its way to its maximum displacement from equilibrium on the other side of its equilibrium position. Then it heads back toward the equilibrium position and passes through it again. This motion repeats itself continually as long as the waves are traveling through the location of the particle of the string in question. The maximum displacement of any particle along the length of the string, from that point's equilibrium position, is called the *amplitude* $y_{max}$ of the wave.

The amplitude can be annotated on both of the two kinds of graphs given above (Displacement versus Position, and Displacement versus Time). Here we annotate it on the Displacement versus Position graph:

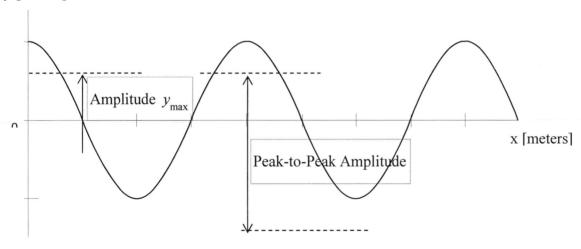

**Figure 13.10**

The peak-to-peak amplitude, a quantity that is often easier to measure than the amplitude itself, has also been annotated on the graph. It should be obvious that the peak-to-peak amplitude is twice the amplitude.

# Period

The amount of time that it takes any one particle along the length of the string to complete one oscillation is called the period $T$. Note that the period is completely determined by the source of the waves. The time it takes for the source of the waves to complete one oscillation is equal to the time it takes for any particle of the string to complete one oscillation. That time is the period of the wave. The period, being an amount of time, can only be annotated on the Displacement vs. Time graph (not on the Displacement vs. Position Along the String graph).

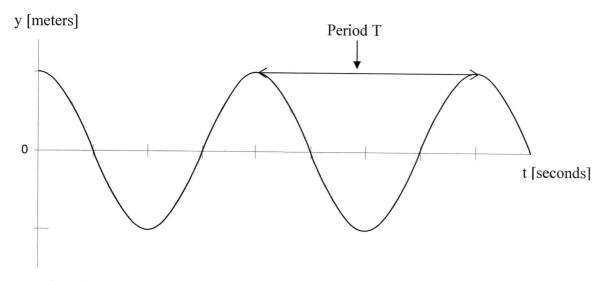

**Figure 13.11**

# Frequency

The frequency $f$ is the number-of-oscillations-per-second that any particle along the length of the string undergoes. It is the oscillation rate. Since it is the number-of-oscillations-per-second and the period is the number-of-seconds-per-oscillation, the frequency $f$ is simply the reciprocal of the period $T$: $f = \dfrac{1}{T}$.

Amplitude, period, and frequency are quantities that you learned about in your study of oscillations. Here, they characterize the oscillations of a point on a string. Despite the fact that the string as a whole is undergoing wave motion, the fact that the point itself, any point along the length of the string, is simply oscillating, means that the definitions of amplitude, period, and frequency are the same as the definitions given in the chapter on oscillations. Thus, our discussion of amplitude, period, and frequency represents a review. Now, however, it is time to move on to something new, a quantity that does not apply to simple harmonic motion but does apply to waves.

# Wavelength

The distance over which the wave pattern repeats itself once, is called the wavelength $\lambda$ of the wave. Because the wavelength is a distance measured along the length of the string, it can be annotated on the Displacement versus Position Along the String graph (but not on the Displacement versus Time graph):

# Wave Velocity

The wave velocity is the speed and direction with which the wave pattern is traveling. (It is NOT the speed with which the particles making up the string are traveling in their up and down motion.) The direction part is straightforward, the wave propagates along the length of the string, away from the cause (something oscillating) of the wave. The wave speed (the constant speed with which the wave propagates) can be expressed in terms of other quantities that we have just discussed.

The wave speed $v$:

$$v = \frac{\lambda}{T}$$

16

One typically sees the formula for the wave speed expressed as

$$v = \lambda f$$

17

where the relation $f = \frac{1}{T}$ between frequency and period has been used to eliminate the period.

Equation ($v = \lambda f$) suggests that the wave speed depends on the frequency and the wavelength. This is not at all the case. Indeed, as far as the wavelength is concerned, it is the other way around—the oscillator that is causing the waves determines the frequency, and the corresponding wavelength depends on the wave speed. The wave speed is predetermined by the characteristics of the string— how taut it is, and how much mass is packed into each millimeter of it. Looking back on our discussion of how oscillations at one end of a taut string result in waves propagating through it, you can probably deduce that the greater the tension in the string, the faster the wave will move along the string. When the hand moves the end of the first segment up, the force exerted on the second segment of the string by the first segment will be greater, the greater the tension in the string. Hence the second segment will experience a greater acceleration. This goes for all the segments down the line. The greater the acceleration, the faster the segments pick up speed and the faster the disturbance, the wave, propagates along the string—that is, the greater the wave speed. We said that the wave speed also depends on the amount of mass packed into each millimeter of the string. This refers to the linear density $\mu$, the mass-per-length, of the string. The greater the mass-per-length, the greater the mass of each segment of the string and the less rapidly the velocity will be changing for a given force. Here we provide, with no proof, the formula for the speed of a wave in a string as a function of the string characteristics, tension $F_T$ and linear mass density $\mu$:

$$v = \sqrt{\frac{F_T}{\mu}}$$

18

Note that this expression agrees with the conclusions that the greater the tension, the greater the wave speed; but the greater the linear mass density, the smaller the wave speed.

## Intensity

Consider a tiny buzzer, suspended in air by a string. Sound waves, caused by the buzzer, travel outward in all away-from-the-buzzer directions:

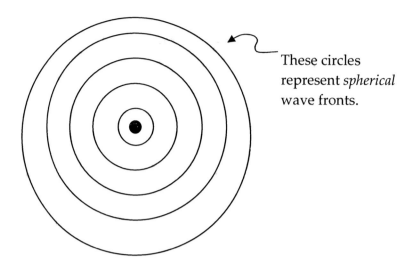

These circles represent *spherical* wave fronts.

In the diagram above, the black dot represents the buzzer, and the circles represent wavefronts— collections of points in space at which the air molecules are at their maximum displacement away from the buzzer. The wavefronts are actually spherical shells. In a 3D model of them, they would be well represented by soap bubbles, one inside the other, all sharing the same center. Note that subsequent to the instant depicted, the air molecules at the location of the wavefront will start moving back toward the buzzer, in their toward and away from the buzzer oscillations, whereas the wavefront itself will move steadily outward from the buzzer as the next layer of air molecules achieves its maximum displacement position and then the next, etc.

Now consider a fixed imaginary spherical shell centered on the buzzer. The power of the wave is the rate at which energy passes through that shell. As mentioned, the power obeys the relation

P $\alpha$ $f^2 y^2 max$

Note that the power does not depend on the size of the spherical shell; all the energy delivered to the air by the buzzer must pass through any spherical shell centered on the buzzer. But the surface of a larger spherical shell is farther from the buzzer and our experience tells us that the further you are from the buzzer the less loud it sounds suggesting that the power delivered to our ear is smaller. So how can the power for a large spherical shell (with its surface far from the source) be the same as it is for a small spherical shell? We can say that as the energy moves away from the source, it spreads out. So by the time it reaches the larger spherical shell, the power passing through, say, any square millimeter of the larger spherical shell is relatively small, but the larger spherical shell has enough more square millimeters for the total power through it to be the same.

Now imagine somebody near the source with their eardrum facing the source.

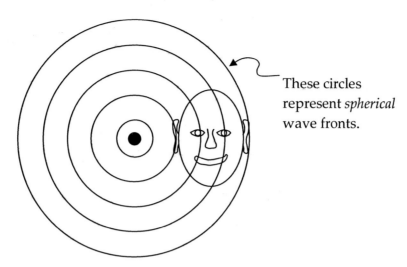

These circles represent *spherical* wave fronts.

The amount of energy delivered to the ear is the power-per-area passing through the imaginary source-centered spherical shell whose surface the ear is on, times the area of the eardrum. Since the spherical shell is small, meaning it has relatively little surface area, and all the power from the source must pass through that spherical shell, the power-per-area at the location of the ear must be relatively large. Multiply that by the fixed area of the eardrum and the power delivered to the eardrum is relatively large.

If the person is farther from the source,

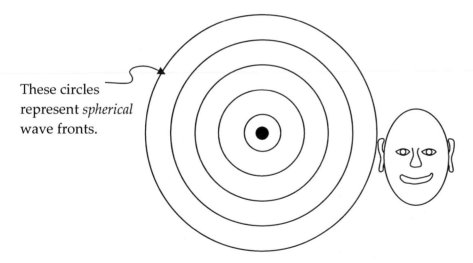

These circles represent *spherical* wave fronts.

the total power from the source is distributed over the surface of a larger spherical shell so the power-per-area is smaller. Multiply that by the fixed area of the eardrum to get the power delivered to the ear. It is clear that the power delivered to the ear will be smaller. The farther the ear is from the source; the smaller is the fraction of the total power of the source, received by the eardrum.

How loud the buzzer sounds to the person is determined by the power delivered to the ear, which, we have noted, depends on the power-per-area at the location of the ear.

The power-per-area of sound waves is given a name. It is called the *intensity* of the sound. For waves in general (rather than just sound waves) we simply call it the intensity of the wave. While the concept of intensity applies to waves from any kind of source, it is particularly easy to calculate in the case of the small buzzer delivering energy uniformly in all directions. For any point in space, we create an imaginary spherical shell, through that point, centered on the buzzer. Then the intensity $I$ at the point (and at any other point on the spherical shell) is simply the power of the source divided by the area of the spherical shell:

$$I = \frac{P}{4\pi r^2}$$
19

where the $r$ is the radius of the imaginary spherical shell, but more importantly, it is the distance of the point at which we wish to know the intensity, from the source. Since P $\alpha$ $f^2 y^2 max$, we have:

I = (amplitude)^2

# Interference

Consider a case in which two waveforms arrive at the same point in a medium at the same time. We wil use idealized waveforms in a string to make our points here. In the case of a string, the only way two waveforms can arrive at the same point in the medium at the same time is for the waveforms to be traveling in opposite directions:

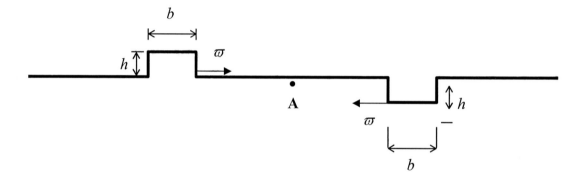

The two waveforms depicted in the diagram above are "scheduled" to arrive at point A at the same time. At that time, based on the waveform on the left alone, point A would have a displacement +h, and based on the waveform on the right alone, point A would have the displacement –h. So, what is the actual displacement of point A when both waveforms are at point A at the same time? To answer that, you simply add the would-be single-waveform displacements together algebraically (taking the sign into account). One does this point-for-point over the entire length of the string for any given instant in time. In the following series of diagrams we show the point-for-point addition of displacements for several instants in time.

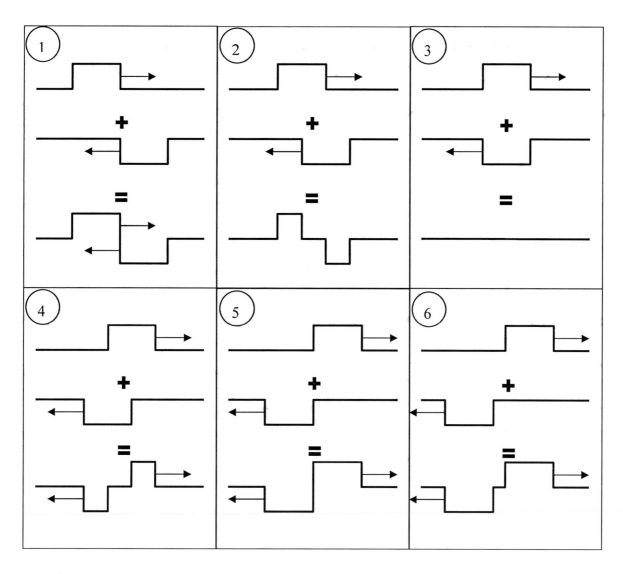

The phenomenon in which waves traveling in different directions simultaneously arrive at one and the same point in the wave medium is referred to as *interference*. When the waveforms add together to yield a bigger waveform,

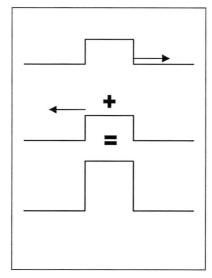

the interference is referred to as *constructive* interference. When the two waveforms tend to cancel each other out,

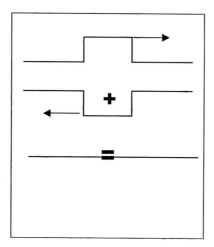

the interference is referred to as *destructive* interference.

## Reflection of a Wave from the End of a Medium

Upon reflection from the fixed end of a string, the displacement of the points on a traveling waveform is reversed.

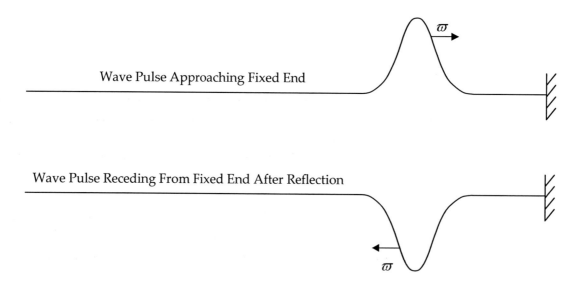

Wave Pulse Approaching Fixed End

Wave Pulse Receding From Fixed End After Reflection

The fixed end, by definition, never undergoes any displacement.

Now we consider a free end. A fixed end is a natural feature of a taut string. A free end, on the other hand, an idealization, is at best an approximation in the case of a taut string. We approximate a free end in a physical string by means of a drastic and abrupt change in linear density. Consider a rope, one of which is attached to the wave source, and the other end of which is attached to one end of a piece of thin, but strong, fishing line. Assume that the fishing line extends through some large distance to a fixed point so that the whole system of rope plus fishing line is taut. A wave traveling along the rope, upon encountering the end of the rope attached to the thin fishing line, behaves approximately as if it has encountered a free end of a taut rope.

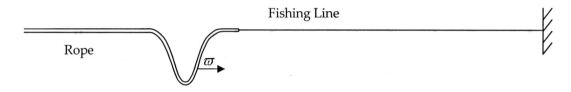

In the case of sound waves in a pipe, a free end can be approximated by an open end of the pipe.

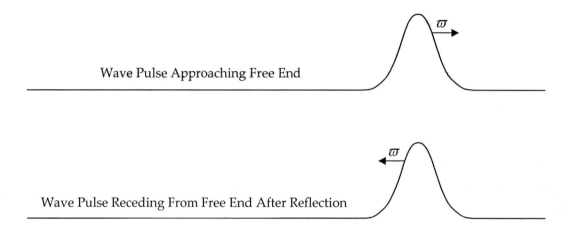

# Standing Waves

Consider a piece of string, fixed at both ends, into which waves have been introduced. The configuration is rich with interference. A wave traveling toward one end of the string reflects off the fixed end and interferes with the waves that were trailing it. Then it reflects off the other end and interferes with them again. This is true for every wave and it repeats itself continuously. For any given length, linear density, and tension of the string, there are certain frequencies for which, at, at least one point on the string, the interference is always constructive. When this is the case, the oscillations at that point (or those points) on the string are maximal and the string is said to have *standing waves* in it. Again, standing waves result from the interference of the reflected waves with the transmitted waves and with each other. A point on the string at which the interference is always constructive is called an *antinode*. Any string in which standing waves exist also has at least one point at which the interference is always destructive. Such a point on the string does not move from its equilibrium position. Such a point on the string is called a *node*.

# Strings, Air Columns

One can determine the wavelengths of standing waves in a straightforward manner and obtain the frequencies from

$V = f\lambda$

where the wave speed V is determined by the tension and linear mass density of the string. The method depends on the boundary conditions—the conditions at the ends of the wave medium. (The wave medium is the substance [string, air, water, etc.] through which the wave is traveling. The wave medium is what is "waving.") Consider the case of waves in a string. A fixed end forces there to be a node at that end because the end of the string cannot move. (A node is a point on the string at which the interference is always destructive, resulting in no oscillations. An antinode is a point at which the interference is always constructive, resulting in maximal oscillations.) A free end forces there to be an antinode at that end because at a free end the wave reflects back on itself without phase reversal (a crest reflects as a crest and a trough reflects as a trough) so at a free end you have one and the same part of the wave traveling in both directions along the string. The wavelength condition for standing waves is that the wave must "fit" in the string segment in a manner consistent with the boundary conditions. For a string of length L fixed at both ends, we can meet the boundary conditions if half a wavelength is equal to the length of the string.

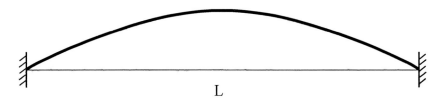

L

Such a wave "fits" the string in the sense that whenever a zero-displacement part of the wave is aligned with one fixed end of the string another zero-displacement part of the wave is aligned with the other fixed end of the string.

Since half a wavelength fits in the string segment we have:

$$\frac{1}{2}\lambda = L$$

$$\lambda = 2L$$

Given the wave speed V, the frequency can be solved for as follows:

$$V = f\lambda$$

$$F = v/\lambda$$

$$F = v/2L$$

It should be noted that despite the fact that the wave is called a standing wave and the fact that it is typically depicted at an instant in time when an antinode on the string is at its maximum displacement from its equilibrium position, all parts of the string (except the nodes) do oscillate about their equilibrium position.

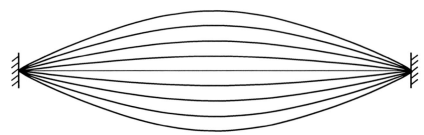

Note that, while the interference at the antinode, the point in the middle of the string in the case at hand, is always as constructive as possible, that does not mean that the string at that point is always at maximum displacement. At times, at that location, there is indeed a crest interfering with a crest. At other times, there is a zero displacement part of the wave interfering with a zero-displacement part of the wave, a trough interfering with a trough, or an intermediate-displacement part of the wave interfering with the same intermediate-displacement part of the wave traveling in the opposite direction. All of this corresponds to the antinode oscillating about its equilibrium position.

The $\lambda = 2L$ wave is not the only wave that will fit in the string. It is, however, the longest wavelength standing wave possible and hence is referred to as *the fundamental*. There is an entire sequence of standing waves. They are named: the fundamental, the first overtone, the second overtone, the third overtone, etc, in order of decreasing wavelength, and hence, increasing frequency.

Fundamental

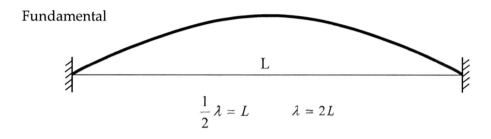

$$\frac{1}{2}\lambda = L \qquad \lambda = 2L$$

1st Overtone

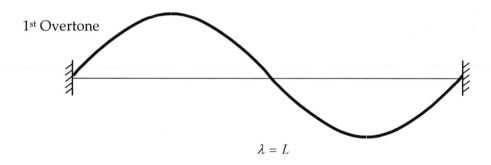

$$\lambda = L$$

2nd Overtone

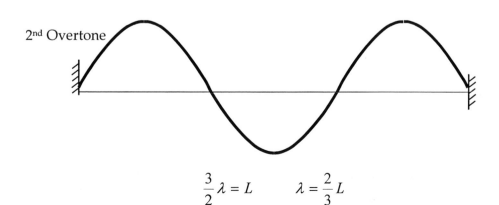

$$\frac{3}{2}\lambda = L \qquad \lambda = \frac{2}{3}L$$

Each successive waveform can be obtained from the preceding one by including one more node.

A wave in the series is said to be a harmonic if its frequency can be expressed as an integer times the fundamental frequency. The value of the integer determines which harmonic (1st, 2nd, 3rd, etc.) the wave is. The frequency of the fundamental wave is, of course, 1 times itself. The number 1 is an integer so the fundamental is a harmonic. It is the 1st harmonic.

Starting with the wavelengths in the series of diagrams above, we have, for the frequencies, using $V = f\lambda$

which can be rearranged to read

$f = v/\lambda$

## The Fundamental

$$\lambda_{FUND} = 2L$$

$$f_{FUND} = v / \lambda_{fund}$$

$$f_{FUND} = \frac{v}{2L}$$

The common formula is can be derived by considering the number of overtone (n) as,

**F (overtone, n)** $= n\dfrac{v}{2L}$

Note that the fundamental is (as always) the 1st harmonic; the 1st overtone is the 2nd harmonic; and the 2nd overtone is the 3rd harmonic. While it is true for the case of a string that is fixed at both ends (the system we have been analyzing), it is *not always* true that the set of all overtones plus fundamental includes all the harmonics.

You may analyze and extend this pattern to other types of wave propagations to air columns too!

# PowerPoint Link:

Please refer to the end of the module lecture link.

# Discussion Question

Answer the following questions with references. Please remember to follow the standard APA referencing style.

For APA standards of references, please visit: http://owl.english.purdue.edu/owl/resource/560/01/

Also, respond in detail to one other post by fellow students.

**13.1** Visit the following website: http://www.youtube.com/watch?v=3mclp9QmCGs

Watch the video and discuss the cause of the disaster, in terms of waves, vibrations, and resonance. Elaborate the effects with relevant equations and formulae.

**13.2** Derive the equations to find the position, velocity, and acceleration in simple harmonic motions.

# Laboratory Activity and the link

Go to the link below and run the simulation:

http://phet.colorado.edu/simulations/sims.php?sim=Pendulum_Lab

Go to: http://phet.colorado.edu/en/contributions/view/3119

Open 2 Student directions Pendulum find g.doc

Do the activity.

Now take the chapter 13 test (not included with this book)

# Module 8

## CHAPTER-14

# Sound

## Objectives

At the end of this lesson, you should be able to:

1.  Describe properties of sound waves

2.  Explain amplitude and frequency

3.  Apply Doppler Effect to calculate changing frequency

## Lecture Notes

### Beats and the Doppler Effect

*Beats*

Consider two sound sources, in the vicinity of each other, each producing sound waves at its own single frequency. Any point in the air-filled region of space around the sources will receive sound waves from both the sources. The amplitude of the sound at any position in space will be the amplitude of the sum of the displacements of the two waves at that point. This amplitude will vary because the interference will alternate between constructive interference and destructive interference. Suppose the two frequencies do not differ by much. Consider the displacements at a particular point in space. Let us start at an instant when two sound wave crests are arriving at that point, one from each source. At that instant, the waves are interfering constructively, resulting in a large total amplitude. If your ear were at that location, you would find the sound relatively loud. Let us mark the passage of time by means of the shorter period, the period of the higher-frequency waves. One period after the instant just discussed, the next crest (call it the second crest) from the higher-frequency source is at the point in question, but the peak of the next crest from the lower-frequency source is not there yet.

Rather than a crest interfering with a crest, we have a crest interfering with an intermediate-displacement part of the wave. The interference is still constructive but not to the degree that it was. When the third crest from the higher-frequency source arrives, the corresponding crest from the lower-frequency source is even farther behind. Eventually, a crest from the higher-frequency source is arriving at the point in question at the same time as a trough from the lower-frequency source. At that instant in time, the interference is as destructive as it gets. If your ear were at the point in question, you would find the sound to be inaudible or of very low volume. Then the trough from the lower-frequency source starts "falling behind" until, eventually a crest from the higher-frequency source is interfering with the crest preceding the corresponding crest from the lower-frequency source and the interference is again as constructive as possible.

To a person whose ear is at a location at which waves from both sources exist, the sound gets loud, soft, loud, soft, etc. The frequency with which the loudness pattern repeats itself is called the beat frequency. Experimentally, we can determine the beat frequency by timing how long it takes for the sound to get loud $N$ times and then dividing that time by $N$ (where $N$ is an arbitrary integer chosen by the experimenter—the bigger the $N$ the more precise the result). This gives the beat period. Taking the reciprocal of the beat period yields the beat frequency.

The beat frequency is to be contrasted with the ordinary frequency of the waves. In sound, we hear the beat frequency as the rate at which the loudness of the sound varies whereas we hear the ordinary frequency of the waves as the pitch of the sound.

## Derivation of the Beat Frequency Formula

Consider sound from two different sources impinging on one point, call it point P, in air-occupied space. Assume that one source has a shorter period $T_{\text{SHORT}}$ and hence a higher frequency $f_{\text{HIGH}}$ than the other (which has period and frequency $T_{\text{LONG}}$ and $f_{\text{LOW}}$ respectively). The plan here is to express the beat frequency in terms of the frequencies of the sources—we get there by relating the periods to each other. As in our conceptual discussion, let's start at an instant when a crest from each source is at point P. When, after an amount of time $T_{\text{SHORT}}$ passes, the next crest from the shorter-period source arrives, the corresponding crest from the longer-period source won't arrive for an amount of time $\Delta T = T_{\text{LONG}} - T_{\text{SHORT}}$. In fact, with the arrival of each successive short-period crest, the corresponding long-period crest is another $\Delta T$ behind. Eventually, after some number $n$ of short periods, the long-period crest will arrive a full long period $T_{\text{LONG}}$ after the corresponding short-period crest arrives.

$$n\Delta T = T_{\text{LONG}} \qquad\qquad 1$$

This means that as the short-period crest arrives, the long-period crest that precedes the corresponding long-period crest is arriving. This results in constructive interference (loud sound). The time it takes, starting when the interference is maximally constructive, for the interference to again become maximally constructive is the beat period

$$T_{\text{BEAT}} = nT_{\text{SHORT}} \qquad\qquad 2$$

Let us use equation to eliminate the $n$ in this expression. Solving equation **Error! Reference source not found.** for $n$ we find that

$$n = \frac{T_{\text{LONG}}}{\Delta T}$$

Substituting this into equation **Error! Reference source not found.** yields

$$T_{\text{BEAT}} = \frac{T_{\text{LONG}}}{\Delta T} T_{\text{SHORT}}$$

$\Delta T$ is just $T_{\text{LONG}} - T_{\text{SHORT}}$

$$T_{\text{BEAT}} = \frac{T_{\text{LONG}}}{T_{\text{LONG}} - T_{\text{SHORT}}} T_{\text{SHORT}}$$

$$T_{\text{BEAT}} = \frac{T_{\text{LONG}} T_{\text{SHORT}}}{T_{\text{LONG}} - T_{\text{SHORT}}}$$

Dividing top and bottom by the product $T_{\text{LONG}} T_{\text{SHORT}}$ yields

$$T_{\text{BEAT}} = \frac{1}{\dfrac{1}{T_{\text{SHORT}}} - \dfrac{1}{T_{\text{LONG}}}}$$

Taking the reciprocal of both sides results in

$$\frac{1}{T_{\text{BEAT}}} = \frac{1}{T_{\text{SHORT}}} - \frac{1}{T_{\text{LONG}}}$$

Now we use the frequency-period relation $f = \dfrac{1}{T}$ to replace each reciprocal period with its corresponding frequency. This yields:

$$f_{\text{BEAT}} = f_{\text{HIGH}} - f_{\text{LOW}}$$

for the beat frequency in terms of the frequencies of the two sources.

## The Doppler Effect

Consider a single-frequency sound source and a receiver. The source is something oscillating. It produces sound waves. They travel through air, at speed $\omega$, the speed of sound in air, to the receiver and cause some part of the receiver to oscillate. (For instance, if the receiver is your ear, the sound waves cause your eardrum to oscillate.) If the receiver and the source are at rest relative to the air, then the received frequency is the same as the source frequency.

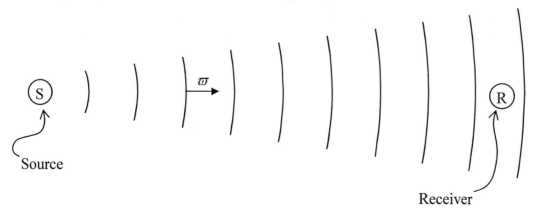

Source

Receiver

But if the source is moving toward or away from the receiver, and/or the receiver is moving toward or away from the source, the received frequency will be different from the source frequency. Suppose for instance, the receiver is moving toward the source with speed $\varpi_R$.

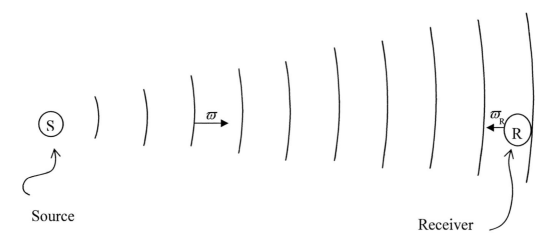

Source                                    Receiver

The receiver meets wave crests more frequently than it would if it were still. Starting at an instant when a wavefront is at the receiver, the receiver and the next wavefront are coming together at the rate $v + V_R$ (where V is the speed of sound in air). The distance between the wavefronts is just the wavelength $\lambda$ which is related to the source frequency(f) $v = \lambda f$ . From the fact that, in the case of constant velocity, distance is just speed times time, we have:

$$\lambda = (v + v_R)T'$$

$$T' = \frac{\lambda}{v + v_R} \qquad\qquad 3$$

for the period of the received oscillations. Using $T' = \frac{1}{f'}$ and $\lambda = v/f$ equation 3 can be written as:

$$\frac{1}{f'} = \frac{v/f}{v + v_R}$$

$$\frac{1}{f'} = \frac{v}{v + v_R}\frac{1}{f}$$

$$f' = \frac{v + v_R}{v}f \quad \textit{(Receiver Approaching Source)}$$

This equation states that the received frequency $f'$ is a factor times the source frequency. The expression $v + v_R$ is the speed at which the sound wave in air and the receiver are approaching each other. If the receiver is moving away from the source at speed $v_R$ , the speed at which the sound waves are "catching up with" the receiver is $v - v_R$ and our expression for the received frequency becomes

$$f' = \frac{v - v_R}{v}f \quad \textit{(Receiver Receding from Source)}$$

Now consider the case in which the source is moving toward the receiver.

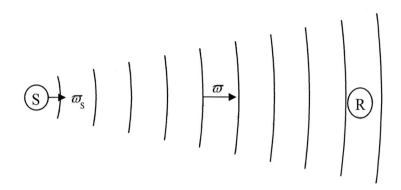

The source produces a crest which moves toward the receiver at the speed of sound. But the source moves along behind that crest so the next crest it produces is closer to the first crest than it would be if the source was at rest. This is true of all the subsequent crests as well. They are all closer together than they would be if the source was at rest. This means that the wavelength of the sound waves traveling in the direction of the source is reduced relative to what the wavelength would be if the source was at rest.

The distance $d$ that the source travels toward the receiver in the time from the emission of one crest to the emission of the next crest, that is in period $T$ of the source oscillations, is

$$d = v_S T$$

where $v_S$ is the speed of the source. The wavelength is what the wavelength would be ($\lambda$) if the source was at rest, minus the distance $d = v_S T$ that the source travels in one period

$$\lambda' = \lambda - d$$
$$\lambda' = \lambda - v_S T \tag{4}$$

Now we'll use $v = \lambda f$ solved for wavelength $\lambda = \dfrac{v}{f}$ to eliminate the wavelengths and $f = 1/T$ solved for the period $T = 1/f$ to eliminate the period. With these substitutions, equation **Error! Reference source not found.**becomes

$$\frac{v}{f'} = \frac{v}{f} - v_S \frac{1}{f}$$

$$\frac{v}{f'} = \frac{1}{f}(v - v_S)$$

$$\frac{f'}{v} = f \frac{1}{v - v_S}$$

$$f' = \frac{v}{v - v_S} f \qquad \textit{(Source Approaching Receiver)}$$

If the source is moving away from the receiver, the sign in front of the speed of the source is reversed meaning that

$$f' = \frac{v}{v + v_S} f \qquad \textit{(Source Receding from Receiver)}$$

The four expressions for the received frequency as a function of the source frequency are combined on your formula sheet where they are written as:

$$f' = \frac{v \pm v_R}{v \mp v_S} f \qquad\qquad 5$$

In solving a Doppler Effect problem, rather than copying this expression directly from your formula sheet, you need to be able to pick out the actual formula that you need. For instance, if the receiver is not moving relative to the air you should omit the $\pm v_R$ . If the source is not moving relative to the air, you need to omit the $\mp v_S$ .

To get the formula just right you need to recognize that when either the source is moving toward the receiver or the receiver is moving toward the source, the Doppler-shifted received frequency is higher (and you need to recognize that when either is moving away from the other, the Doppler-shifted received frequency is lower). You also need enough mathematical savvy to know which sign to choose to make the received frequency $f'$ come out right.

# PowerPoint Link:

Please refer to the end of the module lecture link.

# Discussion Question

Answer the following questions with references. Please remember to follow the standard APA referencing style.

For APA standards of references, please visit: http://owl.english.purdue.edu/owl/resource/560/01/

Also, respond in detail to one other post by fellow students.

14.1 Visit the following website:

http://www.oceanmammalinst.org/songs.html

Listen to the whales' songs and explain the effects of sound waves in air and water by bringing examples with formulae. Discuss the possible changes of whales' songs because of temperature, wave speeds etc. How could you distinguish the whales' songs from other songs?

14.2 Describe the Doppler Effect and derive all related equations.

# Laboratory Activity and the link

Go to the link below and run the simulation:

http://phet.colorado.edu/simulations/sims.php?sim=Pendulum_Lab

Go to: http://phet.colorado.edu/en/contributions/view/3118

Open 1 Student directions pendulum.doc

Do the activity.

# Module 8 Lecture Link:

http://ocw.mit.edu/courses/physics/8-01-physics-i-classical-mechanics-fall-1999/video-lectures/lecture-31/

# Module 8 – Student's Self-Assessment

Please answer the following questions, which give you an indication of your standard of learning for this module.

- Could you describe SHM?

- Could you derive the equation for SHM?

- Can you apply knowledge of SHM to solve problems?

- Can you describe different type so waves?

- Could you describe sound waves?

- Could you calculate amplitude and frequency?

- Can you apply knowledge of Doppler Effect to find frequencies?

- What sections of the textbook would you read for more information on the above?

- Did you enjoy your lesson?

- What aspect of the lesson was most interesting to you?

- Can you now confidently do the lesson's activities?

Now take the chapter 14 test (not included with this book)

You may now take the final test (not included with this book)

# Supplemental Problems and Solutions for selected chapters

## Supplemental Problems for Chapter 1

1. Multiply: $10.5 \times 4.4 \times 6.28$ Express your answer using preferred practices for significant figures.

   *Solution*: When multiplying, the answer is 290.136. However, the least number of significant figures is 2 for the 4.4. Therefore the solution is 290

2. Calculate: $(0.73 + 0.029) \times (3.2 \times 10^3)$. Express your answer using preferred practices for significant figures.

   *Solution*: First perform addition within the parenthesis, and then multiply. The answer is 2428.8. However, the least number of significant figures is 2 for the $3.2 \times 10^3$. Therefore the solution is 2400

3. Find the number of seconds in a 30 day month.

   *Solution*: $30 \text{ days} \times \dfrac{24 \text{ hours}}{1 \text{ day}} \times \dfrac{60 \text{ min}}{1 \text{ hour}} \times \dfrac{60 \text{ sec}}{1 \text{ min}} = 2.592 \times 10^6 \text{ seconds}$

4. If a cement truck can pour 22 cubic yards of cement in one hour, how many cubic feet per minute can be poured?

   *Solution*: $\dfrac{22 \text{ yd}^3}{1 \text{ hour}} \times \left(\dfrac{3 \text{ ft}}{1 \text{ yd}}\right)^3 \left(\dfrac{1 \text{ hour}}{60 \text{ min}}\right) = 9.9 \dfrac{\text{ft}^3}{\text{min}}$

5. A box has dimensions of 0.18 meters by 0.31 meters by 0.090 meters. If there are 3.28 feet per meter, find the volume of the box in cubic feet.

   *Solution*: $0.18 \text{ m} \times 0.31 \text{ m} \times 0.090 \text{ m} \left(\dfrac{3.28 \text{ ft}}{1 \text{ m}}\right)^3 = 0.18 \text{ ft}^3$

## Supplemental Problems for Chapter 2

1. Marcello throws a rock down with speed $12 \dfrac{\text{m}}{\text{s}}$ from the top of a 20 m tower. Assuming the gravitational constant is $9.8 \dfrac{\text{m}}{\text{s}^2}$ and negligible air resistance. What is the rock's speed just as it hits the ground?

*Solution:*  $v_f^2 = v_i^2 + 2as$  Assume that the downward direction is negative. Substituting the initial velocity $v_i = -12 \frac{m}{s}$, acceleration $a = -9.8 \frac{m}{s^2}$ and $s = -20$ m.

$$v_f^2 = \left(-12\frac{m}{s}\right)^2 + 2\left(-9.8\frac{m}{s^2}\right)(-20m)$$

$$v_f^2 = 536\frac{m^2}{s^2}$$

$v_f = 23\frac{m}{s}$  in the downward direction

2.  A race car on a straight section of track accelerates from rest for 5.0 seconds while undergoing a displacement of 49 m.  Find the race car's acceleration if its value may be assumed constant.

*Solution:* $s = v_i t + 0.5at^2$

49 m = 0 (5 sec) + 0.5 a (5 sec)²

$$\frac{49}{12.5}\frac{m}{s^2} = a$$

$a = 3.9 \frac{m}{s^2}$

3.  A train travelling along a straight section of track from Florence to Pisa at a velocity of 78.0 $\frac{m}{s}$ for 1,000 m and then at 60.0 $\frac{m}{s}$ for 900 m.  What is the average velocity?

*Solution:* For the 1,000 m at a velocity of 78.0 $\frac{m}{s}$, the time is found by taking the distance and dividing by the speed which results in 12.8 seconds.  For the 900 m at a velocity of 60.0 $\frac{m}{s}$, the time is 15.0 seconds.

$$\text{Average Velocity} = \frac{\Delta \text{position}}{\text{time}} = \frac{1,000 + 900}{12.8 + 15.0}\frac{m}{s} = 68.3\frac{m}{s}$$

4.  A driver of an automobile traveling at 28.0 $\frac{m}{s}$ sees the traffic signal ahead turn red and applies the brakes bringing the vehicle to a stop in 10.0 seconds.  How far does the automobile travel?

*Solution:* $s = 0.5t(v_f - v_i)$

$$s = 0.5(10s)\left(0 - 28.0\frac{m}{s}\right)$$

s = 140 m

5.  A truck moves 80 m east, then moves 200 m west, and finally moves east again a distance of 110 m.  If east is chosen as the positive direction, what is the truck's resultant displacement?

*Solution*: +80 + (-200) + 110 = -10 m (which means the truck is 10 m west of the starting point)

# Supplemental Problems for Chapter 3

1.  A boat travels 30 miles eastward, then 35 miles northward, and finally 15 miles westward.  What is the magnitude of the boat's net displacement in miles?

    *Solution*:  First make a sketch.

    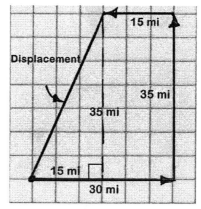

    **Next** calculate the distance, using right triangle trigonometry.

    $d^2 = 15^2 + 35^2$

    d = 38 miles with 2 significant figures.

    A plane is traveling due north, directly towards its destination with an air speed of 250 mph.  A constant breeze is due east at 25 mph.  In which direction is the plane pointed?

    *Solution*:

    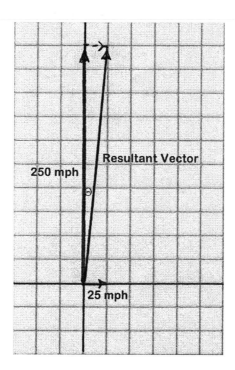

Notice how the vector for the plane only has a y-component and the breeze only has an x-component. In this case, the resultant vector for the plane can easily be found.

$$r = \sqrt{250^2 + 25^2} = 251 \text{ mph}$$

The angle $\theta$ can be found using $\tan \theta = \dfrac{25}{250}$. This results in an angle, $\theta$ of $5.7°$, which is usually expressed as $5.7°$ east of north.

3.  A stone is thrown at an angle of 30° above the horizontal from the top edge of a cliff with an initial speed of 11 $\dfrac{m}{s}$. A stop watch measures the stone's trajectory time from top of cliff to bottom to be 5.8 seconds. (Assume g = 9.8 $\dfrac{m}{s^2}$ and air resistance is negligible.)

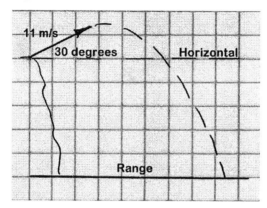

4.  How far out from the cliff's edge does the stone travel horizontally?

    *Solution*: The horizontal velocity of the stone does not change since there is no acceleration in the x-direction. So the distance from the bottom of the cliff to where the stone lands (range) will simply be the x-component of the initial velocity multiplied by the time.

    $$v_{x(\text{initial})} = 11 \cos 30° \ \frac{m}{s} = 9.53 \ \frac{m}{s}$$

    $$\text{Range} = 9.53 \ \frac{m}{s} \ (5.8 \text{ s}) = 55.27 \text{ m} \text{ which is } \textbf{55 m}$$

5.  How high is the cliff?

    *Solution*: In order to find the height of the cliff, the final velocity of the rock must be determined. There is an acceleration in the y-direction due to gravity. Consider the upward direction to be positive; and the downward direction is negative.

    $$v_{y(\text{final})} = v_{y(\text{initial})} + at$$

    $$v_{y(\text{final})} = 11 \sin 30° + (-9.8 \frac{m}{s^2})(5.8 s) = -51.34 \frac{m}{s} \quad \text{(Notice that the velocity is negative indicating a downward direction.)}$$

    $$v_{y(\text{final})}^2 = v_{y(\text{initial})}^2 - 2g\Delta y$$

$$\left(-51.34\frac{m}{s}\right)^2 = \left(11\sin 30°\right)^2 - 2\left(-9.8\frac{m}{s^2}\right)\Delta y$$

Solving for $\Delta y = 132.9$ m or **133m.**

6. A boat is crossing a wide river heading due north with a velocity of 8.0 $\frac{km}{hr}$ relative to the water. The river has a uniform velocity of 4.0 $\frac{km}{hr}$ due east. Determine the velocity of the boat with respect to an observer on the land.

*Solution*: With relative velocity you must use the reference frame of the observer. In this case let us call the velocity of the boat with respect to the river $v_{br}$ and the velocity of the boat with respect to the earth $v_{be}$. The velocity of the river with respect to the earth is $v_{re}$. So the resulting equation for relative velocity will be $v_{br} = v_{be} - v_{re}$.

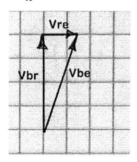

Remember velocities are vectors.

$v_{br}$ has no x-component. $v_{be}$ has an unknown x-component and $v_{re} = 4.0 \frac{km}{hr}$ in the x-direction.

Therefore $0 = v_{be(x)} - 4.0 \frac{km}{hr}$. So $v_{be(x)} = 4.0 \frac{km}{hr}$

$v_{br} = 8.0 \frac{km}{hr}$ is a y-component

$v_{be}$ also has a y-component which is unknown.

The relative velocity equation in the y-direction becomes 8.0 $\frac{km}{hr} = v_{be(y)} - 0$, so $v_{be(y)}$ is 8.0 $\frac{km}{hr}$.

With both of the components vbe can now be found by $\sqrt{4.0^2 + 8.0^2} \frac{km}{hr} = 8.9 \frac{km}{hr}$

## Supplemental Problems for Chapter 4

1.  An elevator weighing 25,000 N is supported by a steel cable. What is the tension in the cable when the elevator is being accelerated upward at a rate of 3.5 $\frac{m}{s^2}$? (g = 9.80 $\frac{m}{s^2}$)

    *Solution*: First make a diagram showing the weight acting downward and the tension in the cable acting upward. Next, there are only forces in the y-direction, so the sum of the forces in the y-direction must be equal to ma.

    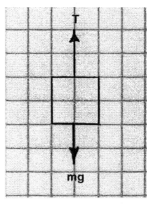

    weight of elevator = mg

    25,000 N = m (9.80 $\frac{m}{s^2}$)

    m = 2,551 kg is the mass of the elevator

    Summing forces: T – mg = ma

    T – 25,000 N = (2,551 kg) (3.5 $\frac{m}{s^2}$) Solving for T results in a cable tension of 33,929 N.

2.  A 2,000 kg sailboat experiences an eastward force of 3,500 N by the ocean tide and a wind force against its sails with magnitude of 6,000 N directed toward the northwest (45° N or W). What is the resultant acceleration and direction of the acceleration?

    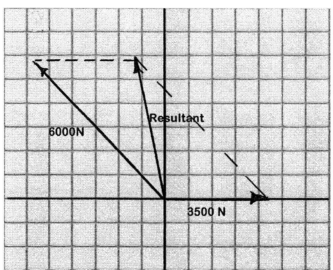

*Solution*: First find the overall resultant force on the sailboat.

Forces in the x-direction : 3,500 N – 6,000 cos 45° resulting in -742.64 N

Forces in the y-direction: 6000 sin 45° resulting in 4242.64 N

The overall resultant force is $\sqrt{(-742.64\text{N})^2 + (4242.64\text{N})^2} = 4307.14\text{N}$ or 4307 N

Notice this force has a negative x-direction and a positive y-direction, so it will be in the northwest quadrant. The acceleration can be found by F = ma, so 4307 N = 2,000 kg (a), resulting in a = 2.15 $\frac{\text{m}}{\text{s}^2}$. To find the angle use $\tan\theta = \frac{4242.64}{-742.64}$, so θ = 80.1° north of west.

3.  A block of mass 6.00 kg rests on a horizontal surface where the coefficient of kinetic friction between the two is 0.25. A string attached to the block is pulled horizontally, resulting in a 2.00 $\frac{\text{m}}{\text{s}^2}$ acceleration by the block. Find the tension in the string.

$(g = 9.80\ \frac{\text{m}}{\text{s}^2})$

*Solution*: Make a diagram as shown below. Remember the friction force acts against movement and is equal to the normal force multiplied by the coefficient of kinetic friction.

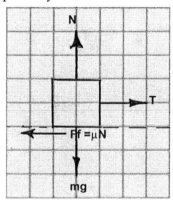

Sum the forces in the y-direction to get the normal force.

$$\Sigma F_y = N - mg = 0 \qquad\qquad \frac{\text{m}}{\text{s}^2}) = 0$$

N = 58.8 newtons (N)

Sum the forces in the x-direction to

$$\Sigma F_x = T - F_{\text{friction}} = ma$$

$$T - 0.25\ (58.8\ \text{N}) = 6.00\ \text{kg}\ (2.00\ \frac{\text{m}}{\text{s}^2})$$

Solving for T gives 26.7 N.

4.  A block of mass 1 kg slides down a plane with an angle of inclination of 30° as shown in the first figure below. The speed-time graph of the block is shown in the 2nd figure. PQ denotes the motion

of the block in portion XY while QR denotes that in portion YZ. (Portions XY and YZ are made of different materials).

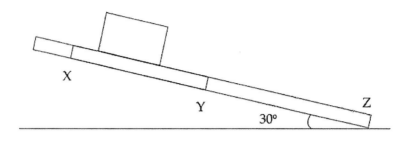

Speed/m/s

(a)

    (i)   Indicate the forces acting on the block in the first figure.

    (ii)  The weight of the block is the resultant force of two force components; one is parallel to the slope of the inclined plane and the other is perpendicular to the slope of the inclined plane. By means of a scale diagram, determine the magnitude of these two force components.

    (iii) Hence, determine the frictional force acting on the block in the portion XY of the inclined plane.

(b)  Determine from the graph in the 2nd figure

    (i)   the acceleration of

    (ii)  the distance travelled by the block in the portion YZ of the inclined plane.

(c)  If the block is projected upwards from Z along the inclined plane with a sufficiently large initial speed (so that the block passes through Y), sketch the speed-time graph of the motion up the plane.

***Solution:***

(a)

   (i)

Reaction Force

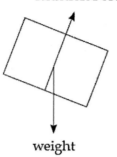

weight

   (ii) Draw to scale with 1 cm to 1 N, with the angles shown below:

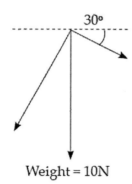

30°

Weight = 10N

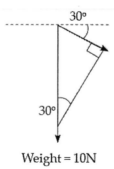

30°

30°

Weight = 10N

        Force component perpendicular to surface = **8.66 N**

        Force component parallel to surface = **5 N**

   (iii) For portion XY, there's zero acceleration as seen from the speed-time graph.

        Hence, net force along surface = 0

        Thus, **frictional force = 5 N**

(b)

   (i) From graph,

       initial speed = 0.75 m/s

final speed = 0 m/s

thus, acceleration = (0 - 0.75) / 2 s

acceleration = **-0.375 m/s²**

(ii)  Distance travelled = area under graph

Distance travelled = ½ (0.75) (2) = **0.75 m**

(c)  It can be seen from part (a) that the friction for portion YZ is greater than for portion XY

Thus, together with gravitational acceleration along the slope, the deceleration for portion YZ will be greater than for portion XY

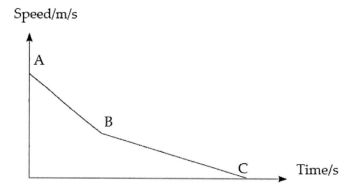

AB represents for portion YZ
BC represents for portion XY

# Supplemental Problems for Chapter 5

1.  What is the kinetic energy of a 0.135 kg baseball thrown at 45.0 m/s?

*Solution:*  $KE = \dfrac{1}{2}mv^2$

$$KE = \frac{1}{2}(0.135kg)\left(45.0\frac{m}{s}\right)^2 = 137 \text{ J}$$

2.  Old Faithful geyser in Yellowstone Park shoots water hourly to a height of 40 m.  With what *velocity* does the water leave the ground?

*Solution*:  Using the Conservation of Energy

$$mgh_i + \frac{1}{2}mv_i^2 = mgh_f + \frac{1}{2}mv_f^2$$

where i is the initial condition and f is the final $m\left(9.8\dfrac{m}{s^2}\right)(40m) = 0 + \dfrac{1}{2}mv^2$  Since the mass is the same on both sides of the equation, cancel m and solve for the velocity.

$v = 28 \dfrac{m}{s}$

3.  Phil pushes a box 5.00 m by applying a 25.0 N horizontal force.  What work does he do?

*Solution*: W = Fd

W = 25.0 N (5.00 m) = 125 J

4. A professional skier reaches a speed of 56 $\frac{m}{s}$ on a 30° slope. Ignoring friction, what was the minimum distance along the slope the skier would have had to travel, starting from rest?

*Solution*: Using the Conservation of Energy

$$mgh_i + \frac{1}{2}mv_i^2 = mgh_f + \frac{1}{2}mv_f^2$$

The initial velocity is 0 which results in the following equation:

$$mgh = \frac{1}{2}mv^2$$

$$m\left(9.8\frac{m}{s^2}\right)h = \frac{1}{2}m\left(56\frac{m}{s}\right)^2$$ Since the mass does not change and is on both sides of the equation,

cancel the m's and solve for h.

h = 160 m which is the height of the slope. In order to find how far the skier must travel, use trigonometry.

$$\sin 30° = \frac{160m}{d}$$

Solving for d yields 320 m.

5. A 2,000 kg car rolls 50.0 m down a frictionless 10.0° incline. If there is a horizontal spring at the end of the incline, what spring constant is required to stop the car in a distance of 1.00 m?

*Solution*: Use the Conservation of Energy

$$mgh_i + \frac{1}{2}mv_i^2 + \frac{1}{2}kx_i^2 = mgh_f + \frac{1}{2}mv_f^2 + \frac{1}{2}kx_f^2$$

Since there is only potential energy initially and only energy due to the car impacting the spring in the final position, the equation simplifies as shown below.

$$(2000kg)\left(9.80\frac{m}{s^2}\right)(8.682m) = \frac{1}{2}k(1.00m)^2$$ where the height of 8.682 m is found by using

trigonometry. $\sin 10° = \frac{h}{50.0}m$

Solving for k gives 340 kN.

To find similar questions and solutions to all chapters, please visit,

http://www.gobookee.net/get_book.php?u=aHR0cDovL3RlYWNoZXJzLnJlZGNsYXkuazEyLmRlLnVz
L3dpbGxpYW0uYmFrZXIvUGh5c2ljyUyMEZpbGVzL3BoeXNpY3MlMjBzb2x1dGlvbnMlMjBtYW51Y
WwlMjBmb3IlMjBwcmFjdGljZS1lbmQlMjBjaGFwLnBkZgpTb2x1dGlvbnMgTWFudWFsIC0gUmVkIE
NsYXkgQ29uc29saWRhdGVkIFNjaG9vbCBEaXN0cmljdA==

This is a free ebook available under http://www.gobookee.net.